Kaizen and Kaizen Event Implementation

Kaizen and Kaizen Event Implementation

Chris A. Ortiz

PRENTICE HALL

Upper Saddle River, NJ • Boston • Indianapolis • San Francisco
New York • Toronto • Montreal • London • Munich • Paris • Madrid
Capetown • Sydney • Tokyo • Singapore • Mexico City

Many of the designations used by manufacturers and sellers to distinguish their products are claimed as trademarks. Where those designations appear in this book, and the publisher was aware of a trademark claim, the designations have been printed with initial capital letters or in all capitals.

The author and publisher have taken care in the preparation of this book, but make no expressed or implied warranty of any kind and assume no responsibility for errors or omissions. No liability is assumed for incidental or consequential damages in connection with or arising out of the use of the information or programs contained herein.

The publisher offers excellent discounts on this book when ordered in quantity for bulk purchases or special sales, which may include electronic versions and/or custom covers and content particular to your business, training goals, marketing focus, and branding interests. For more information, please contact:

U.S. Corporate and Government Sales
(800) 382-3419
corpsales@pearsontechgroup.com

For sales outside the United States please contact:

International Sales
international@pearson.com

Visit us on the Web: informit.com/ph

Library of Congress Cataloging-in-Publication Data

Ortiz, Chris A.
 Kaizen and kaizen event implementation / Chris A. Ortiz.
   p.   cm.
 Includes index.
 ISBN 0-13-158456-1 (pbk. : alk. paper)
 1. Total quality management. 2. Organizational change. I. Title.
 HD62.15.O75 2009
 658.4'013—dc22   2009003555

Copyright © 2009 Pearson Education, Inc.

All rights reserved. Printed in the United States of America. This publication is protected by copyright, and permission must be obtained from the publisher prior to any prohibited reproduction, storage in a retrieval system, or transmission in any form or by any means, electronic, mechanical, photocopying, recording, or likewise. For information regarding permissions, write to:

Pearson Education, Inc.
Rights and Contracts Department
501 Boylston Street, Suite 900
Boston, MA 02116
Fax (617) 671-3447

ISBN-13: 978-0-13-158456-3
ISBN-10:   0-13-158456-1

Text printed in the United States on recycled paper at Courier in Stoughton, Massachusetts.

First printing, April 2009

Contents

Preface xi
Acknowledgments xv
About the Author xvii
Introduction xix

Part I Kaizen Basics 1

1 Kaizen and Kaizen Events 3
  Kaizen 4
    People of Kaizen 5
    Leaders of Kaizen 5
    Benefits of Kaizen 7
  Kaizen Events 8
    Common Mistakes Made in Kaizen Events 9
    Metrics 12

2 The Company Kaizen Program 17
  Kaizen Event Steering Committee 18
    Plant or General Management 19
    Manufacturing Engineering Management 19
    Quality Management 20
    Operations or Production Management 21
    Human Resource Management 21
    Maintenance or Facilities Management 22
    Purchasing or Materials Management 23

　　　　Production Supervisor　23
　　　　Operator Representative　24
　　Introducing the Kaizen Champion　24
　　Tracking　25
　　　　Kaizen Event Selection　26
　　　　Date and Length　28
　　　　Kaizen Event Team Leader　28
　　　　Kaizen Team Members　28
　　　　Preplanning and Preplanning Responsibility　29
　　　　Pre-event Goals　30
　　　　Actual Results　30
　　　　Event Budget and Event Spending　30
　　　　Action Items, Responsibility, and Status　31
　　Kaizen Communication　33
　　　　Kaizen Communication Boards　34
　　　　Kaizen Newsletter　34
　　　　Kaizen Suggestion Box　35

3　The Kaizen Champion　39
　　Why a Kaizen Champion?　40
　　Kaizen Champion Skill Sets　41
　　　　The Seven Wastes　41
　　　　Lean as a Business Model　42
　　　　5S and the Visual Workplace　42
　　　　Kaizen and Kaizen Events　43
　　　　Data Collection　43
　　　　Setup Reduction and Quick Changeover　45
　　　　Line Design and Work Flow　46
　　　　Material Replenishment　46
　　　　Project Management　47
　　Choosing the Kaizen Champion　47
　　　　Internal Option　48
　　　　External Option　48
　　Cost of a Kaizen Champion　49

Kaizen Champion Responsibilities   50
　Training   50
　Kaizen Monthly Meeting   51
　Communication Boards   51
　Kaizen Newsletter   51
　Kaizen Suggestion Box   52
　Kaizen Event Tracking Worksheet   52
　Team Leadership   52
　Action Item Follow-up   52
　Monitoring Other Lean Initiatives   53
Alternatives   53
Why a Kaizen Champion?   54

4 Kaizen Event Scheduling   55
Four Weeks Before the Kaizen Event   56
　Select the Process/Department/Work Area   56
　Make a Tentative List of Kaizen Team Members   57
　Select the Kaizen Event Team Leader   58
　Establish Team Goals   59
　Estimate Event Spending   59
　Order Supplies   60
　Update the Kaizen Communication System   61
　Schedule Outside Assistance   61
　Conduct Waste Analysis of the Area   61
Two Weeks Before the Kaizen Event   62
　Finalize the Kaizen Team Members   62
　Get an Update on Supplies and Outside Resources   63
　Ask Team Members to Walk Through the Chosen Area   63
　Pick a Room Where the Team Can Gather   63
　Analyze the Collected Data and Start Coming Up
　　with Design Ideas   64
One Week Before the Kaizen Event   64
　Gather Current State Information   64
　Meet with the Kaizen Team Members   65
　Place All Supplies in the Team's Gathering Space   65

Meet with the Plant or General Manager   65
Make Food Preparations   66
Final Thoughts on the Timeline   66

Part II Kaizen Events   67

5   5S Kaizen Events   69

Four Weeks to Go   70
   Select the Area   70
   Select the Team Leader   70
   Tentatively Select the Team Members   71
   Establish Goals   71
   Event Spending and Supplies   71
   Update the Kaizen Communication System   72
   Identify the Kaizen Team Meeting Space   72
   Schedule Outside Assistance   72

Two Weeks to Go   72
   Finalize Kaizen Team Members   72
   Get an Update on Supplies and Outside Resources   72
   Ask Team Members to Walk Through the Selected Area   73

One Week to Go   73
   Gather Current State Information   73
   Meet with the Kaizen Team Members   73
   Place All Supplies in the Team's Meeting Room   74
   Meet with the Plant or General Manager   74

Let It Begin!   74
   Day One: Sort   74
   Day Two and Day Three: Set in Order and Scrub   79
   Day Four: Standardize   83
   Day Five   84

Maintenance 5S Events   84
   Day One: Sort   85
   Day Two and Day Three: Set in Order and Scrub   86
   Day Four: Keep Going   87
   Day Five   87

5S Sustaining Tips 87
　　Create an End-of-Day Cleanup Procedure 88
　　Conduct a Daily/Shift Walk-through 88
　　Establish a 5S Audit Sheet 88
　　Create and Maintain a 5S Tracking Sheet 89
　　Develop a 5S Incentive Program 90

6 Standard Work Kaizen Event 91

Preplanning 92
　　Effective Time 92
　　Volume Requirements 92
　　Takt Time 92
　　Process Analysis 93
Day One 98
　　Red-Tag Team 99
　　Review the Line Balancing Information 100
　　Completing the Red-Tag Event 101
　　End-of-Day Meeting 102
Day Two 102
　　Line Design 102
　　Midday Meeting 103
　　Scrub/Shine 104
　　Maintenance and Machine Shop Projects 104
　　Tool Presentation 104
　　Shadow Boards 105
　　Subassemblies 105
Day Three 106
Day Four 106
　　Creating Workstation and Parts Rack Signs 107
　　Floor Taping and Designations 108
　　Determining Subassembly Build Levels 108
　　Installing Shadow Boards and Tower Lights 109
Day Five 110

7 Case Study: Samson Rope Technologies, Inc. 111
January 2007 112
Kaizen and Kaizen Event Implementation Training 113
The Return Visit 114
Lean Assessment and Strategy Sessions 115
Samson Rope's Kaizen Program 116
    Kaizen Steering Committee 116
    Communication 116
    Kaizen Event Supply Box 117
    Monthly Meeting 117
Kaizen Event 1, May 7–11, 2007: Cell 5, Cell 8, and Splicing 117
5S Continued 120
Kaizen Event 2, September 24–28, 2007: Maintenance 120
Completing 5S 122
Kaizen Event 3, December 3–7, 2007: Coating 123
2007 Complete 124
Lafayette, LA: Lean Assessment 124
Kaizen Event 1, February 20–22, 2008: Large Rope 901 125
Kaizen Event 2, April 7–11, 2008: Area G, Area D, Area B 127
Samson Rope Progress: Ferndale and Lafayette 128
Other Samson Rope Employees to Recognize 129

8 Conclusion 131

Index 133

Preface

How long do new ideas, once implemented, last? Many of you may have embarked on lean manufacturing journeys over the years, implementing 5S, single-piece flow, standard work, kanban, or something else, but you find that nothing seems to "stick." Let me illustrate a familiar manufacturing situation.

It is business as usual at the Company A production facility. The manufacturing supervisor is walking back and forth, watching the interactions between people and product like a general assessing his troops before battle. A constant sense of urgency forces workers to ignore standard operating procedures and critical quality responsibilities.

Poorly designed workstations do not adhere to any ergonomic or safety criteria. Work content is severely imbalanced, forcing operators to stand around looking for something to do rather than working. Parts and material presented to each workstation are poorly organized, so operators are walking around searching for items essential to performing their work.

Crisis management and volume are typical preoccupations; people resolve issues as they arise rather than finding long-term solutions. The day ends with the usual production meeting with its semiheated exchange of problems and quick fixes.

Does this sound like your company?

A company I worked for years ago had a similar culture. Resolution efforts were disorganized. Too many people were involved, and they worked with little direction. Small, separate teams were created to focus on various problems and areas simultaneously. There were a lot of them: a continuous improvement team, a process engineering team, a quality engineering team, a 5S team, and a lean engineering team.

Each team was assigned tasks the company deemed important to implementing lean. Each team met regularly to discuss its projects and possible improvements, not knowing what the other groups were addressing.

Team members had to juggle their usual day-to-day responsibilities with the team's assigned improvement efforts. With no clear vision or goals, teams worked for months on end, never resolving anything. The outcome was inconsistent participation, tension, low morale—and, worst of all, no results.

It is pretty clear that a company like this needs better ways of implementing lean manufacturing processes. Yet implementation of lean does not necessarily guarantee success if there is no support from top management. Without that support and a foundation for change, new ideas will fall by the wayside. There has to be focus, dedication, and commitment to making and sustaining long-term improvements.

The philosophy of kaizen, which means continuous improvement, can create that foundation. Here is how a company kaizen program can help a company like Company A:

1. **Create a kaizen committee.**
   Not just another team, this is a group of upper-management leaders who oversee all kaizen-event-related activities. They are in charge of scheduling kaizen events, selecting team members, and keeping everyone accountable for completing projects on time. A kaizen committee can determine the lean initiatives and select team members from the various engineering groups to implement lean on the factory floor.

   With this leadership committee in place, there would be no more misconceptions about what to do and when to do it. With management support, team members could dedicate 100 percent of their time to the kaizen event, and their usual day-to-day responsibilities would be given to others until after the event.

2. **Create a champion.**
   Company A needs an employee who is totally dedicated to driving the continuous improvement efforts in the organization. The only thing on a kaizen champion's mind is lean. He or she can help train employees on the fundamentals of lean and kaizen. The kaizen champion is responsible for monitoring the changes made through kaizen events and keeps employees aware of upcoming events.

3. **Train employees.**
   Employees in Company A, including operators, need to understand fundamental lean and kaizen concepts such as 5S, standard work, visual management, waste reduction, and takt time. As time goes on, more and more employees will begin to understand the importance of lean and kaizen, and a culture of change agents is created.

4. **Hold monthly meetings.**
   Company A's kaizen committee should meet once a month and schedule kaizen events. All kaizen events should be scheduled four weeks in advance for planning purposes. Team members should be selected two weeks in advance, to allow them to plan accordingly and so that their managers can prepare for their absence. Vacation time can be verified to ensure that team members are available for the event. This two-week time frame allows the team members to make arrangements at home if the team is to work an off shift.

5. **Conduct kaizen events.**
   Kaizen events are used to implement continuous improvement on the factory floor. Holding an event every month will help the organization's culture evolve into one of continuous improvement. Before anyone can go back to the old ways of working, another kaizen event is going on. Over time, new standards and procedures will be created and resisting change will become harder.

6. **Develop vision and focus.**
   The keys to implementing lean are vision and focus. Company A had some vision, but no focus. Permanent changes will not happen if a company simply "grazes" along with improvement initiatives. A company can be staffed with the best lean talent, but without the infrastructure to encourage and sustain improvements, the grazing will continue. Developing a company kaizen program to act on this vision and focus is essential to lean success.

Kaizen and Kaizen Event Implementation was designed to serve two purposes. First, it is a valuable book for plant managers and middle managers who are relatively new to lean and are looking for guidelines to create an infrastructure for continuous improvement. Second, it can be used in organizations that have just started their lean journeys and need new ideas to accelerate their programs. Plant managers, engineering managers, lean champions, and even directors can use *Kaizen and Kaizen Event Implementation* to jump-start lean in their organizations.

Kaizen and Kaizen Event Implementation is an essential tool for your company and can be used to develop a powerful and long-lasting continuous improvement journey for all readers. It is my hope that you find tremendous value in this book and will use it as your field guide for implementations for the life of your organization.

Acknowledgments

First and foremost, I thank my family: my wife, Pavlina, and my two sons, Sebastian and Samuel. Without their support and encouragement, it would have been nearly impossible to devote the time to this book that I did. All of my professional efforts are to ensure our success together.

Second, I want to thank the staff at Prentice Hall who first took interest in this book for their professional approach to making it become a reality. I want to acknowledge my editor, Bernard Goodwin, for his passion for lean manufacturing and for taking on my project.

Third, I would like to thank all the employees of Samson Rope Technologies in the Ferndale, WA, and Lafayette, LA, facilities. The relationship we have developed in the two years we have been working together has been professional and highly enjoyable. Thank you for allowing me to use your story in this book.

It is important also to recognize the countless manufacturing professionals I have had the luxury of working with over the years. My clients are the true experts; I learn from them, and the relationships I have developed with them are critical to my company's success.

About the Author

Chris Ortiz is the president and executive lean consultant of Kaizen Assembly. He has spent the majority of his professional career working for Fortune 500 companies, teaching and guiding them to become more efficient. Chris has also led more than 150 kaizen events around the country. His company's clientele includes Samson Rope Technologies, Wood Stone, Hexcel Corporation, Messier-Bugatti, Engineered Solutions, Prince Castle, Bellingham Cold Storage, Absorption Corp., Erin Baker's Wholesome Baked Goods, Trans-Ocean, and IKO Pacific Inc., to name just a few.

Chris is the author of the book *Kaizen Assembly: Designing, Constructing, and Managing a Lean Assembly Line* (CRC Press) and *Lessons from a Lean Consultant: Avoiding Lean Implementation Failure on the Shop Floor* (Prentice Hall). His lean implementation techniques have been featured in a variety of trade magazines, newspapers, and corporate newsletters, such as *Industrial Engineer*, *Industrial Management*, *Process Cleaning Magazine*, and many other trade publications. He is an active speaker at engineering conferences and expos around the country.

Chris Ortiz can be reached by e-mail at chrisortiz@kaizenassembly.com, or go to his Web site, www.kaizenassembly.com.

Introduction

Lean is not a "program"-based concept. This makes understanding lean difficult for many professionals because we all like nice step-by-step guidelines. In its simplest form, lean is about removing waste or non-value-added effort in a company. Removing or reducing this waste is a never-ending battle. By continually focusing on waste reduction, a company can react better to the needs of its customers and also operate at more efficient performance levels. To understand the phenomenon of lean, one must learn the tools within the philosophy and see how they are intertwined. Kaizen is one of those tools. Kaizen and lean are often confused; many people think that kaizen is lean, but it is only a part of the lean philosophy.

The most commonly used lean tools are

- Kaizen
- 5S
- Standard work
- Setup reduction and quick changeover
- Kanban
- Quality at the source
- Total productive maintenance (TPM)

Kaizen

Kaizen is a Japanese word for "continuous improvement." Kaizen involves all employees in a company focusing on process improvements. This first piece of the lean journey is often confused with lean itself, but kaizen is not lean. Lean is about removing waste; kaizen is about continuous improvements. Kaizen is part of lean. This book focuses on kaizen and the use of kaizen (continuous improvement) events.

5S and the Visual Workplace

5S is a methodology of organizing, cleaning, developing, and sustaining a productive work environment. The 5Ss are

- **Sort:** the act of removing all unnecessary items from the work area
- **Straighten:** organizing what is needed so everything has a home and its identity and location are clearly marked
- **Scrub:** cleaning everything
- **Standardize:** maintaining consistency in the visual workplace
- **Sustain:** maintaining improvements and continually improving upon them

A visual workplace is one with no clutter and better visibility of problems so that employees can be more proactive. Items such as tools, parts, documentation, and supplies can be easily located for quicker access. 5S alone is an extremely powerful improvement tool for productivity, quality, and safety, but also for general appearance and increased morale. Like kaizen, 5S is just part of lean.

Standard Work

Another powerful improvement tool is *standard work*. Standard work is essentially "best practice." It is an agreed-upon set of work procedures that establishes the most efficient, most reliable, and safest methods and sequences for each process and each worker. In a standard work environment everyone has clear roles and responsibilities. More important, people and machines are used to their fullest potential, and workloads are evenly spread out. For instance, work content required in each workstation on an assembly line should be outlined in detail, and cycle times should be as even as possible. This allows for better flow, and it places the same workload on each individual. If one operator has 5 minutes' worth of work in a workstation and another one has 3½ minutes, then they are not evenly balanced; either the person with 5 minutes of work is overloaded, or the operator with 3½ minutes of work is underloaded, depending on the flow requirements. The time of the person with 3½ minutes of work is not maximized; hence the company is less productive.

Material handlers should have specific routes and routines and assigned areas of responsibility. A person operating a piece of equipment should follow certain setup and machine-run tasks, and this work should be associated with a time standard. To ensure higher productivity and better use of time, the machine operator needs to follow this standard work. Standard work is supported through the proper documentation of work instructions that outline the requirements of the work. These instructions could be in the form of assembly instructions, setup instructions, changeover instructions, material handling maps and routes, cleanup procedures, and start-up procedures. The list can go on and on. Standard work is an integral part of lean and must be incorporated at some point.

Setup Reduction and Quick Changeover

Reducing setups and the time associated with changeover is an absolute must in a lean environment. Setup by definition is non-value-added. The customer is not willing to pay for the extra time or cost your organization incurs performing it. Changeover is the process of setting up a machine, equipment, or a production line for another process or product. This time is downtime during which no value-added work is being performed. Many factories have excessive setup times when machines or processes are not operating. Companies can create a lot of problems with long setup times.

Excessive work in process (WIP) and finished goods can accumulate. Rather than changing over more frequently, the manufacturer simply builds more than is needed, knowing that the downtime will be significant. This is a "what if" scenario in that there is anticipation of orders or of future need for the part or product. The problem is not output; it is the long changeover time. Extra inventory can add to the internal cost of the organization. The cost of added inventory is not simply the cost of the parts; an entire infrastructure is required to maintain and control it. It requires people, floor space, racks and shelving, software, forklifts, paperwork, and computers. Inventory can get damaged while in stock, or entire lots can be manufactured incorrectly and the problem may not surface for months, when the item is pulled to place an order—when it's too late. So reducing the time associated with setups and changeovers is extremely important in a lean journey.

Kanban

Kanban is a material replenishment system that incorporates signals, pull instructions, visual cues, bins, carts, containers, etc., to help coordinate material and parts transactions throughout the factory and with suppliers. Material and parts are kept in specified quantities and containers when needed. Implementing a kanban system will help your company reduce the amount of inventory and help predict better flow of material. It will help simplify scheduling and improve productivity.

The amount of material or parts for each kanban item depends on product volume, size of the part, lead time from the stockroom, supplier lead times, supplier quantities, market trends, and variations in your model mix. Each kanban system, like all lean systems, must be tailored to each business model and structure. There is no "one size fits all."

Quality at the Source

Quality at the source is an approach to quality that places the responsibility for catching errors in the hands of the operator, or at the point of build. Successful implementation of this lean tool requires a major shift in how supervisors and operators look at quality. Quality is not just the responsibility of the quality control department. The process in which the product is manufactured must be set up to allow production line workers to recognize errors before they become defects. The development of a proactive culture is needed in quality at the source, and a mind-set of error prevention rather than reaction to problems should be taught.

Line workers should be required to perform specified incoming and outgoing checks on every unit or with reasonable frequency. The production line or work area should have the proper quality-check tools as well. When operators check their work and their coworkers' work, the chances of a defect occurring diminish rapidly. Once the product reaches a midline or final inspection point, it should pass with no issues. It will take some time to change the mind-set of the workers who may have just relied on end-of-line checks to ensure quality. In this lean system, quality is everyone's job.

Total Productive Maintenance (TPM)

If you have a factory that uses machines and equipment, TPM will be a very valuable lean tool. TPM is a preventive maintenance (PM) approach that creates employee ownership and encompasses proactive machine upkeep. First of all, machines, tools, and fixtures should be set up for fast changeovers, easy operations, and preventive maintenance. You have to train your operators properly in setup, changeover, and running the equipment, including the necessary safety and cleanup procedures. Avoid purchasing cheap and potentially unreliable equipment. Although there is a short-term cost savings, in the long run the initial savings will be lost to downtime, poor quality, and missed delivery dates, and the equipment could jeopardize worker safety.

A TPM program should have three levels. The first level is the TPM required by the machine operator. These tasks include daily cleanup and checking operating conditions such as fluid levels, heat, and power. First-level TPM is relatively simple and should be performed daily. The second level of TPM is the work performed by a maintenance department less frequently, maybe once a week or month, depending on the machine's use. Sometimes second-level TPM requires a total or partial teardown of the machine for repairs or replacing parts. The third level of TPM is the work performed by the manufacturer of the machine. This may be done once or twice a year. All levels of this TPM program are worth the investment in time and money to ensure that your equipment can operate at optimum productive levels and last the length of its life cycle.

The Seven Wastes

The purpose of a kaizen event is actually quite simple: to remove or reduce waste. I say *reduce* because there is no such thing as a waste-free workplace. With what has already been mentioned in this introduction, you can probably put some of the pieces together. The concepts discussed are implemented to reduce waste, and can be done through kaizen events or any other kaizen-related activities. Let's go over the seven wastes, as they are the focal point of all improvement initiatives and the reason this book was written.

- **Overproduction:** The act of producing more than is needed, before it is needed, and faster than is necessary. Overproduction is by far the most common type of waste in an organization, and it can breed other wastes.

- **Overprocessing:** This occurs when it is hard to see when something is complete. For instance, grinding, sanding, and polishing can be overdone, because a sense of completion is hard to gauge from one person to the next. Redundant effort or steps, and excessive checking and verifying, are examples of overprocessing. If operators need to unpackage parts from suppliers on the production line before installing those parts, they are overprocessing.

- **Motion:** Unnecessary movement of people in the plant or in the general work area, such as looking for parts and tools, leaving the work area for any reason, and physically moving products and parts. Motion is probably the second most common waste.

- **Waiting:** When manufacturing and operational process are out of synchronization, people and machines are idle.

- **Transportation:** Movement of material (raw material, WIP, and finished goods).

- **Inventory:** Excessive levels of raw material, WIP, and finished goods in correlation to throughput time and delivery requirements.

- **Defect/rework:** Quality errors that have become costly and were not prevented.

In the course of your lean journey, you will learn of other lean philosophies and tools that can be used to reduce waste.

How you use and mix these tools depends on your culture, company, and processes. More important, they are all simply part of the lean philosophy. This book is dedicated to teaching you about kaizen and kaizen events. The first chapter will outline a company's struggle with how lean is applied in its organization. As employees juggle multiple projects, deal with day-to-day issues, and are asked to wear many hats, finding time for lean is difficult. The chapter will then dive into the fine detail of kaizen and kaizen events, comparing the two and showing you common mistakes made in developing a kaizen program.

The concept of a company kaizen program is then described in Chapter 2. Topics will include the kaizen event steering committee and a kaizen champion who is 100 percent dedicated to continuous improvement. It will also discuss tracking and scheduling kaizen events as well as kaizen communication. This chapter will give you information on how to piece together the program that will embrace ongoing change.

The next chapter is entirely dedicated to discussing the kaizen champion. A kaizen champion is essentially the lean torchbearer and is in place to help drive all lean and kaizen initiatives. The selection or recruitment of this person should not be undertaken lightly. It is a high-profile position, and the person in it can make the difference between having a moderately successful lean journey or a very successful one.

Chapter 4 is dedicated to kaizen event preparation, including event timelines, team selection, team leader selection, team objectives, and kaizen event supplies.

The first four chapters lay the groundwork for the remainder of the book. Chapters 5 and 6 will describe how to use kaizen events for the implementation of 5S, standard work, kanban, and a new line design. They will help you see how kaizen events are used for implementation and ongoing improvements to your organization. Each kaizen event concludes with a formal presentation, called a report out, and a tour. It is important to invite as many employees in the company as possible to this report-out session so that the team can discuss their accomplishments and how they improved the performance and culture of the business. Chapter 6 will conclude with this information.

Finally, Chapter 7 is dedicated to a case study from a company called Samson Rope Technologies located in Ferndale, WA, and Lafayette, LA, which used the information from this book to begin a lean journey.

Part I
Kaizen Basics

The first part of this book is intended to outline the fundamental aspects of a company kaizen program. I will describe the difference between kaizen as a philosophy and kaizen events, the implementation mechanism for many lean initiatives. The kaizen champion, the kaizen steering committee, communication, meetings, the kaizen event supply box, and the important steps in planning events will all be discussed.

one

Kaizen and Kaizen Events

When companies make the decision to embark on a lean journey, they frequently have a few misconceptions about the endeavor. First of all, some believe that lean is a program with definable starting and ending criteria. In addition, this "program" is supposed to have clear direction and paths that dictate what to do and when to do it. If lean is incorporated with this mind-set, the chances of failure are very high. The concepts of lean and kaizen are incorporated into business in a manner that is right for the company. If I compared how each of my clients has adopted lean, I would find one definable similarity: They *started*. Once a lean journey begins, who knows where it will go? In lean, there is no fixed path or one-for-all guideline.

I am not implying that lean journeys do not involve setting goals for improvement such as increasing productivity, reducing scrap, improving on-time delivery, reducing inventory, or decreasing throughput time, for example; but how each company works to accomplish these types of metrics is different. You cannot adopt one organization's practices and apply them to your own organization in exactly the same way. I often see this confusion when teaching the "phenomenon" of lean. People struggle to connect the dots and see how it will work in their organization. It is this first misconception I would like to discuss in this chapter.

Kaizen

Kaizen is a Japanese word for "continuous improvement and incremental change." The philosophy of kaizen is about involving everyone in the organization to focus on overall organizational improvements. The cornerstone of lean manufacturing is removing waste to better respond to the needs of the customer in regard to on-time delivery, competitive cost, and better quality. More important, kaizen emphasizes developing a process-oriented culture that is driven to improve the way a company operates. Think of the number of processes that exist in a company. A process generally has a starting point and an ending point. To clarify, the process of manufacturing and assembling a product starts with fabricating and processing parts from raw material; then those parts are fitted together to make the final product. This is a simple and crude example, but my point is that the process by which these products are built ends at some point, or else there would be nothing tangible remaining. Let's apply this concept to an administrative/office environment. There is a process by which a purchase order is created or a contract is generated. Both processes have a start and an end, when the purchase is completed and sent to the warehouse or production floor, or when the contract is signed by both parties.

By removing waste, an organization becomes more productive, ensuring that it is serving the customer's needs. This will bring a financial gain to the organization, but you cannot sell lean to a culture if you are only promoting its cost savings. Let's be honest; reduced cost, better quality, and on-time delivery will not encourage all employees to change the way they think. The philosophy of kaizen brings much more to the table. Changing company culture is an ongoing battle, and you want to address issues that may arise early on. So in essence, kaizen is about coaching and mentoring people to become better at what they do in all aspects of their work. Buying expensive pieces of equipment or software will not bring the cultural change you need to make lean successful. These types of expenditures usually create a one-time improvement with minimal effort. By no means does this imply that there is no need for capital expenditures, but kaizen does not mean spending a lot of money.

So to refer back to what I previously wrote, there is no perfect road map for dealing with company culture, and it is this culture that will determine your level of success and distinguish your company from other organizations.

People of Kaizen

Some companies place the responsibility for process improvements on manufacturing engineers and managers. These individuals generally come up with the initial ideas to improve a work area, conduct the analysis and preplanning, and then implement the change. Production workers feel that the new process is being "pushed" on them because they were not given the opportunity to suggest improvements. This is not the case across all spectrums, but it is still a very common practice. In kaizen-based organizations, process improvements involve everyone from executive leadership down to the entry-level production worker. This includes the creation of the improvement idea, process analysis, preparation phases, implementation, and training. The kaizen philosophy not only encourages production workers to suggest improvements but requires that they do so. This can be difficult for some leaders to swallow because it essentially means relinquishing some of their authority in the improvement process. I have come across many plant and other upper managers who find it difficult to delegate decision making for the company. The most successful lean journeys occur, however, when upper and even executive managers back off and provide an environment that fosters change. When people are allowed to speak openly and make changes from their own perspectives, the possibilities are endless. Managers who allow and encourage this behavior will see far more progress in their organization's lean journey than those who tend to make all the decisions themselves.

Leaders of Kaizen

How do kaizen and lean fit into a company's vision? A common illusion that business leaders have is that lean is the one and only business strategy for the company. Lean is indeed a business strategy, but it should not be the all-encompassing focus. As organizations develop their overall vision and focus, lean has to be a major role player. Again, the concepts of lean and kaizen, when all is said and done, deliver value to the customer in terms of cost, quality, and delivery. A company's culture must be driven to continuous improvement because it benefits the customer. Delivering that value is difficult because each customer is different and expectations are always changing. Lean transformation is one tool for achieving better customer relations, but there are other tools in a company's strategy, such as improving supplier relations, training and mentoring employees, adding product lines, and capturing new markets and business segments. These are examples of possible "pillars" that would be part of a larger strategy. One of those pillars is lean/kaizen.

Becoming a leader of kaizen takes time because leaders are part of company culture just like engineers, maintenance personnel, and production workers. Transformation into a kaizen leader does not happen overnight. As I mentioned in the preceding section, kaizen leaders must learn to release some of their hold on authority and give it to everyone in the company so that change and improvements can spread through the organization. Next, kaizen leaders must not focus on the financial gain from lean but rather on using kaizen to help develop their people.

In my previous book, *Lessons from a Lean Consultant*, I wrote an entire chapter called Lean Leadership Made Simple. The mentality of company leaders who practice negative management techniques—working their people long hours and using them as cogs in the wheel—is devastating to the lean journey. Allow me to summarize from that chapter.

My personal experiences in the lean field have taught me a lot of valuable things, especially how to treat people. The companies I have assisted quickly realized that a new approach to leadership was needed to ensure success in their lean endeavors. I was by no means a perfect employee in the years leading up to starting Kaizen Assembly, and in fact I was a bit resistant to lean as well. However, I always maintained the belief that my resistance was normal and appreciated my great lean leaders. How we treat people in our lean journeys is the cornerstone of lean leadership.

I took all that I learned from my experiences and use it now to lead companies in a manner that seems fair and just. Organizations embarking on lean need effective leaders who understand the importance of employee contributions and how much their efforts and attitudes affect the success or failure of a company. Certain corporate leaders need to realize that although aggressive practices may result in short-term financial success, they also place the company on the path toward a precarious future.

Lean leaders are only human beings; therefore, they typically conduct themselves in a manner that reflects their personality. If individuals are generally grumpy and negative to change, their management techniques will reflect those characteristics, and they will affect the morale of others through their body language as well as their words. Individuals who are happy and positive tend to lead in the same manner. Lean leaders who do not let negativity influence their actions will create a following of positive thinkers.

Management techniques reflecting personalities can be categorized in the following ways. Poor lean leadership definitely results in lack of motivation, poor performance, high absenteeism, and, ultimately, high employee turnover. Poor lean leaders are easily recognizable because they have some or all of the following characteristics: They are focused on their own personal needs rather than the professional needs of their team; they are pessimistic rather than positive; they are poor listeners; they are lazy or lack motivation; they are stubborn or closed to new ideas; they are slow to adapt to change; they are blamers rather than responsibility takers; they provide bad or unclear direction; they have no idea who their people are; they are secretive; they are never available; their doors are always closed; they fear failure; they do not stand behind their people; they have difficulty developing their employees; they exercise leadership by control, manipulation, and coercion. None of these qualities is helpful in successfully engaging people in lean.

Effective lean leadership is not based on control, coercion, and manipulation. Lean leaders are focused on the future rather than the past. They gain respect by their ability to inspire others to work toward specific goals. Effective lean leaders help others to become better people; they create workplaces that attract good individuals, and they keep their workers happy, motivated to pursue excellence, and focused on continuous improvement.

Kaizen is simply a mind-set and philosophy of ongoing change and improvement. As a lean practitioner I am often asked how to deal with resistance to change. There is no perfect template or guideline for dealing with people. You and your company have to work continually with your employees and provide the support and accountability they need to mold them into your own change agents.

Benefits of Kaizen

Kaizen teams are created to provide a quick and positive impact on the organization. Each team member is handpicked according to his or her ability to make both measurable and nonmeasurable improvements. Kaizen events teach people the concepts of teamwork, meeting deadlines, interacting with different personalities, and pursuing excellence as a whole, and they open up employees' creativity. Professional and personal relationships are developed during kaizen events that continue after the events are over. These are examples of nonmeasurable benefits that allow the organization to develop a culture driven toward continuous improvement. The other side to kaizen events is more measurable: Teams make improvements to key metrics that not only benefit the company from a performance perspective, but ultimately improve the relationship with the customer in regard to better cost, on-time delivery, and improved quality.

Kaizen Events

Learning the theory behind kaizen is important as you begin your lean journey. Now let's talk about how to turn that philosophy into action. Often called a *rapid improvement project*, a kaizen event is a set time frame that is scheduled to allow a group of employees to come to together and implement lean and remove waste. The core of this book outlines how to create a company kaizen program and, more importantly, how to schedule, conduct, and follow up on kaizen events.

Kaizen events are structured time-wise and are very project-based. However, companies can get into a mode where they only wait for the kaizen events to make improvements. This is called *event-lean*. Kaizen events allow for the "shock and awe" effect and can positively impact company performance, but the test of an organization's ability to keep the momentum going is to identify waste removal opportunities in between kaizen events.

Ideally, a company should try to get to the point where it can conduct kaizen events every month. Don't expect to do this in the first year. Maybe scheduling kaizen events once a quarter or every other month is best in the beginning. It depends on your culture, production schedules, and what other important projects and activities are going on in the company. My job in this book is to provide you with information that will allow you to schedule monthly kaizen events. As time goes on, you will become better at planning and conducting them.

Many organizations use kaizen events but still cannot create a culture that embraces change, and many improvement efforts fall short of their cultural and financial goals. The reason behind this is that the company did not have the infrastructure in place to keep everyone involved, motivated, and, more important, wanting more. Kaizen events can become annoying to some if the events are disorganized and under management that does not support the efforts. Management must set clear direction on why kaizen events are important and place specific goals in front of each team. I will outline these important ingredients of the program in this book.

Common Mistakes Made in Kaizen Events

Kaizen events require focus and solid up-front planning. A company will have to allocate resources and invest time and money in the program. Kaizen and kaizen events do not require a lot of money, but in all honesty, money will be spent. However, the rate of return will be phenomenal. Kaizen teams also need goals placed before them to provide challenge and excitement. Never walk into a kaizen event without team goals at some level. Here are some common mistakes made in planning and conducting kaizen events:

- Lack of communication
- Lack of planning
- Poor team selection
- No goals

Lack of Communication

Communication will be addressed in greater detail in Chapter 3, but allow me to describe it a little here. The mistake that organizations make is not communicating to all employees that lean and kaizen are going to be a way of life. As kaizen events are scheduled, they must be made known and their importance understood. Everyone should know when the events are to take place, who is on the team, who the team leader is, what area has been selected, and the goals and objectives for the team. This way the factory will know who will be relieved of their normal responsibilities to focus on the kaizen event. Ongoing communication about lean projects shows commitment from leadership, and that kaizen events will not go away.

A couple of years ago I was discussing a potential partnership with the manager of a plant that manufactures lighting fixtures. This organization had been conducting kaizen events and was three years into its lean journey. The conversation led to the importance of ongoing events and communication. The manager believed in basically saturating his people with information about kaizen events, and regardless of other organizational activities, an improvement project was always scheduled. In his words, "hell or high water, we are having a kaizen event." Before the culture could revert to established routines and business as usual, another kaizen event was coming. Ongoing information about the progress of the lean journey is essential to keeping the lean fire lit.

Lack of Planning

Solid up-front planning is critical to the success of kaizen events. When I was a young industrial engineer learning about lean, I came into contact with many consultants and trainers. Some of the earlier kaizen event teachings did not emphasize the need for preplanning. The leader was supposed to walk into day one of a kaizen event and do the training and analysis on that day. Very little preparation was done, and as I used this philosophy, I saw how it negatively affected the results of the project. Some kaizen events require very little preparation and others involve prior analysis. In Chapter 4 I will break down the key tasks that should be completed prior to kaizen events, starting at four weeks ahead all the way down to the day before the event.

The number of preplanning activities will vary depending on the kaizen event. Analyzing waste, conducting time studies or process mapping, and analyzing flow may be necessary to establish a current state. Supplies such as floor tape, bins, racks, signs, paint, and labels may need to be ordered. Maybe the company wants to bring in employees from sister plants or from suppliers to be part of the event. Tools and equipment might need to be reserved or rented. It is important to think about these things ahead of time to ensure that everything is ready for the event.

Poor Team Selection

Selecting the right employees to participate in a kaizen event—gathering a good mix of talents and disciplines—is the single most important aspect of any event. As previously mentioned in this chapter, lean and kaizen involve everyone, so your kaizen teams will be different every time. A mistake that is often made when putting together kaizen teams is not selecting people from the production line. Operators and frontline personnel possess intimate knowledge of the process and product, and creating their early buy-in is a key ingredient to sustaining improvements. Each team will need a maintenance person, line operators, engineers, managers, material handlers (if applicable), and maybe another office employee. The number of team members will depend on the complexity of the kaizen event and what needs to be accomplished; I will discuss this in greater detail in Chapter 3. By having a diverse kaizen team, you enable the group to come up with a greater variety of improvement ideas than would probably come from a team consisting of just managers and engineers.

No Goals

A company kaizen program is established and in place to act as a foundation for making and maintaining improvements. Part of this foundation is creating goals. Kaizen teams need clear goals and objectives from the company. The ultimate goal of lean is to fulfill the needs of the customer. Outside of developing positive and healthy relationships, the customer's expectations come in the form of the three business drivers: cost, quality, and delivery. The job of an organization is to find the competitive balance among the three. If an organization is meeting or exceeding the expectations of on-time delivery, the cost of products and services, and the quality of services for its customers, then it is focusing on the right things. So how is lean connected to cost, quality, and delivery? And with respect to kaizen team goals, what type of metrics should be improved? Figure 1-1 illustrates the connection between improving metrics and improving the customer's expectations. All of the metrics can negatively or positively affect cost, quality, and delivery. The goals for each kaizen team should focus on at least two of these metrics since they are directly connected to the customer.

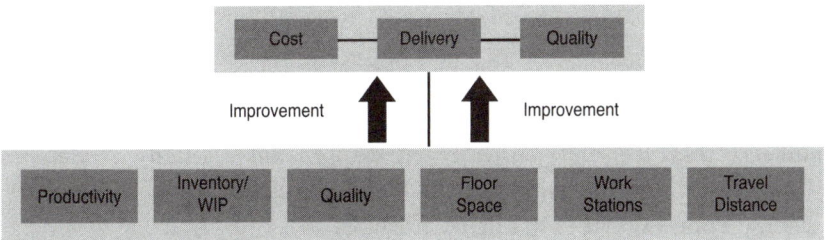

Figure 1-1 Metric connection

Metrics

Productivity

Productivity is improved when more products are made and more services are provided with less effort. The less material, parts, manpower, time, utilities, paperwork, processes, and steps that are needed, the more costs can be reduced. Quality is improved because the opportunities for error are decreased. Fewer steps mean faster throughput and better delivery. Kaizen teams should always pursue productivity improvement goals.

Inventory/WIP

There are essentially three types of material: raw material, partially finished goods, and 100 percent finished goods. Sheet metal, the raw material, can be cut into small pieces such as brackets, plates, or covers. These parts then become work in process (WIP) and can be placed into a product and moved on to another processing step. Partially completed products move through various stages of assembly or processing until they become a finished product ready for sale. Regardless of the stage that this material is in, it costs money. Inventory should be kept at a minimum throughout the plant regardless of its phase in manufacturing. Manufacturing processes should be short with minimal steps so that unnecessary WIP does not build up. WIP can hide quality errors that may eventually lead to rework. Kaizen teams can focus their efforts on reducing inventory levels and WIP. Obviously a lower level of inventory can reduce cost, but it can also improve quality by creating better visibility of problems that can potentially hide in excess WIP. And large clumps of WIP are stopping points or slow-moving points that can adversely affect delivery.

Quality

Improving quality is essential to maintaining and acquiring customers. The last thing you need is to be known as a supplier of poor products and services. I feel that people in general are loyal to quality over anything else. A small percentage is looking for the cheapest deal, but when it comes down to it, quality prevails.

Quality starts with the culture an organization has developed, particularly a mind-set of proactive error prevention rather than reactively dealing with problems. Errors will occur if human beings are part of the manufacturing process. Even in highly automated environments, machines and equipment require human interaction such as setup, maintenance, programming, cleanup, and changeover. An automated process's output is only as good as the human input.

The concept of quality at the source is an effective lean approach to quality that places the responsibility for checking and rechecking the product at the point of build. Frontline production workers need to check the product at various stages of manufacturing to ensure that errors are being caught. Errors are cheap; defects are not. Operators must perform certain incoming and outgoing checks throughout the process. They should check work done in the previous process, or by a previous worker, then perform their own task, and then perform a quality check on the work they just did. Quality at the source results in a tremendous improvement in overall quality. When checks are performed throughout the process, multiple eyes are on the product. This results in a product that is virtually error-free by the time it reaches a more formalized inspection point at the end of the line. Self- and successive checks are very common in a lean journey, but only correct implementation of these checks will ensure that they are performed. Kaizen team goals for quality could be to reduce scrap costs, rework hours, and testing errors, for example. These measures will definitely reduce costs, the end product will be of better quality, and with fewer mistakes and rework, promised delivery dates are attainable.

Floor Space

Manufacturing companies often overuse their existing floor space as unneeded items begin to accumulate. Also, processes themselves are too long and too wide, which results in longer paths for parts and products. Over time, less space is available for production and growth. Sometimes manufacturers come up with plans to physically expand the existing building to accommodate new product lines and products. I say you should "lean it out first before adding." Kaizen teams can focus on reducing the amount of floor space being used for current processes. Once space is better used, new product implementations or capacity increases can take place. When floor space use is reduced, cost is reduced because companies begin to take proactive approaches to buying items for the production floor, purchasing only when something is really needed. Quality is improved because less clutter means less chance for part damage. Better use of floor space means smaller and simpler processes that will help in meeting delivery requirements. Fewer complications, less distance, and fewer physical obstacles equal on-time delivery.

Workstations

Depending on the type of operation, companies may use a traditional manual assembly line with workstations. If you are a machine shop, you could have computer numerical control (CNC) machines, mills, drill presses, and other types of computer-controlled equipment. Maybe your process includes work areas where production workers simply have a place to work. Regardless of the type of work area on the factory floor, the right numbers of people, machines, and stations are essential for better performance. Sometimes it is simply a question of creating a better ratio of people to machines, more efficient equipment utilization, or increased uptime. That right mix needs to be effectively associated with demand.

Station reduction or better use of stations goes hand in hand with floor space use and productivity. Having fewer stations means less "stuff"—fewer workbenches, parts, shelves, tools, paperwork, fixtures, lights, etc. Using the appropriate number of workstations limits the number of people in the process and therefore decreases the opportunity for error. Again, I am not implying job loss, just smarter use of people and the work they perform. Kaizen teams can have a goal of reducing the number of workstations, consolidating processes, or coming up with a more balanced workload among operators. Reducing the number of workstations reduces the cost associated with extra items and too much labor. Quality is improved and work content among workers is balanced and better defined, so work areas can be better used. Fewer stations and processes required to complete products mean faster delivery.

Travel Distance

Longer than needed processes generate plenty of waste. Longer production lines and part flow paths require more people, extend lead times, and add inventory. Travel distance is 100 percent linked to delivery. It takes more time for something to travel 300 feet than 30 feet. The longer a product is in the building, the more money it costs. As an example, I led a kaizen event where the team was required to reduce travel distance by 30 percent. It was a reasonable goal, and the team focused their waste reduction efforts on achieving that goal. After they calculated the correct number of workstations for the assembly line, balanced the work between stations, and converted to single-piece flow, travel distance went from 350 feet to 50 feet. By eliminating 300 feet of travel distance, the team reduced the throughput time by 82 percent, from 11 hours to 2 hours. Think about the customers waiting for their products on these lines . . . delivery, delivery, delivery!

So, kaizen teams can have travel distance reduction goals based on the respective product line. Cost is reduced simply because it requires less effort to complete the product. Quality is improved because there is less distance to travel and fewer chances for error. And delivery—well, I think I have said enough about that.

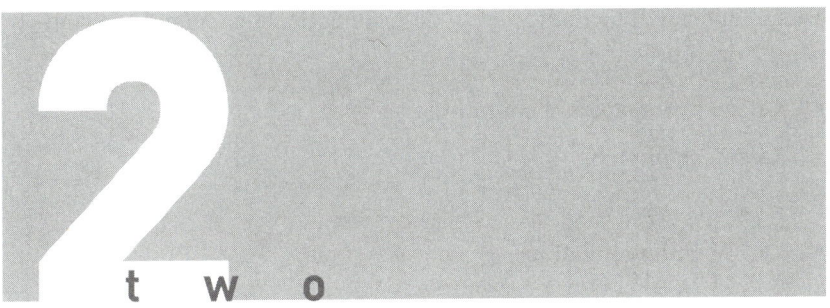

The Company Kaizen Program

A company kaizen program acts as the organizational policy for lean and continuous improvement. Like any other operational policy, this program is in place to ensure that improvements are made and that management is behind the efforts. In the early stages of a lean manufacturing journey, companies are faced with a variety of challenges and obstacles that impede how effectively and quickly new lean processes are implemented.

Manufacturing engineers, as an example, have a collection of responsibilities that require them to deal with day-to-day problems, write procedures, update bills of materials, and train production personnel. Managers are even busier, as they have authority over multiple people, attend countless meetings, and participate in decision-making processes. And let's not forget the floor supervisors who run the production lines and processes every day, interacting with operators and line leads, putting out fires, and managing hour-by-hour crises. When do they all have time to "do lean"?

How can lean be implemented in a manner that is organized, smart, and effective? Often missing from a company's lean manufacturing journey is a foundation that embraces continuous improvement. This foundation is a company kaizen program.

This chapter will be dedicated to describing the elements of the company kaizen program:

- Kaizen event steering committee
- Kaizen champion
- Tracking
- Kaizen communication

I will then address how to customize the program for your organization.

Kaizen Event Steering Committee

The first element of the kaizen program that should be implemented is a governing body of key decision makers in the company who schedule and watch over all lean- and kaizen-related activities. This committee provides the required resources and time to the rest of the employees to ensure that progress is being made in the lean journey and that kaizen teams are supported. The kaizen event steering committee is responsible for a variety of tasks and initiatives, and it is their support that will help slowly change and engage the culture in continuous improvement.

Organizational structures vary in size and in the titles they use for different positions, so there is no perfect committee structure that can be used in every company. However, certain disciplines and areas of responsibility need to be accounted for on the committee. So before I can outline the requirements of the committee, here is a list of titles and why the people who have these titles should be members. Again, look at why each position is on the committee and decide who in your organization can fulfill that role.

- Plant or general management
- Manufacturing engineering management
- Quality management
- Operations or production management
- Human resource management
- Maintenance or facilities management
- Purchasing or materials management

- Production supervisor
- Operator representative

Plant or General Management

Most manufacturers have an individual who watches over the entire plant floor. Often called a plant manager or general manager, this person should be part of the kaizen event steering committee. First and foremost, the plant manager is one of the key decision makers in the company as well as an influence in setting goals for the company and driving lean from a "captain's" perspective—basically, one of the original lean vision setters. The plant manager will approve budgets for the kaizen program and each kaizen event. Each member of the committee essentially reports to the plant manager, so this person is there to keep all upper and middle managers on the committee accountable for providing the appropriate people and help for each kaizen event. However, the plant manager does not make the final call on what is involved in the lean journey; the committee as a whole makes that decision.

All lean projects and initiatives being conducted throughout each year must fall in line with your company's lean strategy. This lean strategy is based on measurable results that were discussed in Chapter 1; productivity, quality, throughput time, floor space, inventory, and workstation quantity are all metrics that, when improved, can make positive changes to on-time delivery, cost, and overall quality for your customers. A lean strategy outlines the lean goals for the company, and all efforts at continuous improvement should be directed to those goals. Don't forget that the committee will also track and monitor the smaller projects outside of structured kaizen events as well. So the plant manager's job is to ensure that whatever is planned by the kaizen event steering committee is driven by the strategy.

Manufacturing Engineering Management

Engineering departments can be extremely different from one company to the next. Smaller companies may not even have a "department" but may have an individual who provides the appropriate technical support to the manufacturing processes. If you do not have a manager per se, then someone who works in the overall manufacturing process is fine. Manufacturing engineering or even industrial engineering departments and people provide a number of services to the company. These departments are in charge of line layout, equipment optimization, collecting data, making work instructions, troubleshooting on the line, training production workers, and making engineering changes to products. They may provide testing and quality support as well.

The manufacturing engineering managers must be on the committee. Kaizen teams and other smaller, lower-profile teams will be making process improvements that could involve new line configurations, changing work content, modifying testing and inspection procedures, shortening lines, developing new changeover procedures, and many other "process"-oriented modifications. It is essential that the manufacturing engineering manager provide the required resources from his or her department to guide the kaizen team as needed. There may also be some additional support for conducting observations and analysis of the prospective process to be used for improvement. This department can help collect this information and provide it to the kaizen team or whoever needs the data (I will describe data collection in later chapters).

Quality Management

The quality manager can provide a lot of information during the preplanning phase. Kaizen teams will need information about internal and external quality on each product and process. Internal quality data such as rework, scrap, defect rates, first-pass yield, and reject rates is valuable information to gather so that the teams can make improvements to these metrics. Also, information collected from customers can potentially be used. The number of complaints, on-site technical service calls, and warranty claims, for example, can help kaizen and kaizen-event-related efforts. Not all quality issues in the field or with the customers are directly related to the factory, but external data can still play a role in improving quality.

Like all the other managers on the kaizen event steering committee, the quality manager must provide the right people from his or her department to assist the team in developing new work instructions and testing and inspection procedures. The quality manager can also provide adequate support to guide the team in the correct approach to part protection when constructing new workstations or work cells.

Operations or Production Management

The individual in your organizational structure who is responsible for the production workers, line leads, and supervisors at your facility should be on the committee. The success of any improvements made, regardless of how small or large in scale they are, depends on those frontline workers who participate directly in the process. The production manager can help create the vision and address the accountability required to sustain improvements. Production workers and line leads will be the first to resist, and the production manager must be strong and assertive in leading people in the new lean process.

Kaizen events are scheduled in advance so that the managers on the committee can make the appropriate preparations. Kaizen events will have a short-lived negative impact on the prospective process as changes are being made. The production manager can prepare the operators and make adjustments in the schedule as needed. Workers may need to move to a second shift during the kaizen event; overtime may be needed to build up product in preparation for downtime or simply to provide added resources for the process during the event to support the production workers.

Kaizen events will use production workers, and the production manager will have to adjust his or her workforce during the kaizen event. Like all other managers, the production manager is responsible for training employees at some level. As changes are made to the line or process, new training will be needed, and the production manager can allocate the resources to help. The main reason the production manager is on the kaizen event steering committee is that he or she will make or break the improvements that are implemented. The people who report to the production manager must know that their leader is behind the changes 100 percent.

Human Resource Management

The human resources (HR) department plays a vital role in the continuous improvement efforts in the company. The obvious reason is that HR is responsible for all the employees, including the plant manager, and has information about each employee specifically. For instance, during kaizen events, participation must be 100 percent, and as team members are selected, HR can verify vacation schedules. More important, HR should know which employees are on light duty or have other work-related restrictions. Maybe the employees being selected for the kaizen event are already committed to other training during that week or will be out of the plant.

As companies develop a stronger kaizen program, ongoing training will become important. Training of new employees is critical because each new person should be made aware of the importance of lean. HR will have to modify the new-employee orientation program at the company and set up introductory lean and kaizen curriculums. The HR department should be fully engaged in and fully supportive of lean efforts.

Maintenance or Facilities Management

Depending on the needs of the organization for better flow of products, parts, and information, changes to the plant layout and all the production processes may be needed. Either through kaizen events or from gradual changes over time, the maintenance department is extremely important in moving equipment and lines. It can be very time-consuming to disconnect and reconnect air and electrical lines. There may be network drops and the electrical lines could be of varying voltages. The time associated with some of these moves can be long. Your organization is going to be using this department on every kaizen event. The maintenance manager must provide the right person from his or her department to support the kaizen team. More important, as assembly lines, work cells, and equipment are moved around, they all have to be operational for the production workers when the moves are complete.

Maintenance personnel are also very good at designing and building customized shelves, fixtures, toolholders, and other great knickknacks for the workstations. In just about every kaizen event I have seen, the maintenance team member was busy welding, cutting, sawing, sanding, and painting something for the team. Maintenance people are very creative and can come up with great ideas for the presentation of tools, parts, testing devices, lights, and documentation. I realize, however, that not all companies have the luxury of a fully staffed maintenance department with ample welders, drill presses, cutting ability, and other great resources. As you develop a lean structure, you may want to consider having the resources at some level to support these needs.

Purchasing or Materials Management

As physical changes are made to the production process, material and parts will be moved. Some of the material may be big and bulky, requiring pallets to hold it. Some parts may be in larger totes or other containers requiring a forklift. The purchasing manager can allocate the right person with a forklift certification. A kaizen event can quickly come to an end if there is no forklift support.

There may also be a need to remove items from the stockroom that are stored up high. If the team decides to move large equipment around, a forklift is most likely needed. The purchasing or materials manager may be the one kaizen event steering committee member responsible for ordering supplies for the kaizen team. After the area has been selected and goals are established, this person can begin the preplanning work for purchasing.

Production Supervisor

The committee members described so far have all been managers. In years past I recommended that my clients have only managers on the committee. I have developed and changed the company kaizen program over time, and I now suggest that certain nonmanagerial employees be placed on the kaizen committee. Production supervisors bring a wealth of knowledge and experience from the production floor's "front line." Also, when a company is implementing lean and making improvements, production supervisors will be the ones who motivate and encourage the production workers. You can select one production supervisor or rotate supervisors every quarter. This could be a good approach to ensuring that all supervisors are part of the early planning and decision-making process.

A word of caution: When employees enter the kaizen event steering committee meeting (which I will discuss shortly), all titles go out the window. Regardless of who sits on this committee, the group as a whole is making decisions to help the continuous improvement program. Leave other subjects and topics for a later time. This also means, for example, that the production manager cannot treat the production supervisors, who may report to him or her, like subordinates.

Operator Representative

Over time, as I have worked with several companies of varying structures, some have opted to add operators to their committee. Generally the operators selected for the committee are veteran long-term employees who have a lot of knowledge of the company. These individuals basically represent the production workers' needs and speak on their behalf. It may be good to rotate an operator representative every quarter to get a nice mix of ideas and perceptions. Production workers are typically used to being given direction by a supervisor or manager. However, on this committee the operator representative has equal say in the improvement efforts. You will also find that some of the best ideas will come from this operator representative, and as you begin to use production workers from the early stages of idea creation all the way to the sustaining phase, your frontline culture will become more and more excited and engaged in the lean initiatives.

The kaizen event steering committee is a vital part of the lean journey. It is a great outlet for gathering ideas and developing a culture of team players. As the company moves toward an overall approach to making decisions, the lean journey is strengthened and accelerated. With my clients, this committee is created very quickly and early in the process. I encourage you to put together your committee soon after reading this book to get things moving.

Introducing the Kaizen Champion

Embarking on a lean journey does take time and resources. Although lean is a way of thinking, it also requires training, implementation, planning, and constant focus on improvements. Companies quickly begin to see that employees' roles and responsibilities begin to change. The lean effort is a company-wide effort. As your company begins to learn the fundamentals of lean, such as 5S, standard work, setup reduction, and kanban, one or more people will have to lead the charge. This is where someone who is totally dedicated to lean and kaizen can make a world of difference. We call this person a kaizen champion. This is not an industry term, as the title could be lean champion, lean engineer, or lean liaison. It really does not matter; what does not change is the role of this person in the company. Pure and simple, it's 100 percent lean!

Once the kaizen champion has been identified, he or she becomes the leader of the kaizen steering committee and runs all the meetings. It is the job of the champion to ask the committee for assistance and to keep the members accountable for contributing. The kaizen champion should have a high level of authority, being able to go directly to the plant manager to get what is needed. It is a kind of gray position, since the kaizen champion is not a manager but has the "pull" of an upper manager. This type of authority is needed to keep management committed to the lean journey.

The kaizen champion is the "torchbearer" of lean and drives all kaizen-related initiatives. He or she essentially is the director of the lean program. This position is so important that I dedicate an entire chapter (Chapter 3) to it. However, please remember that it is not a requirement to have a champion; some companies cannot justify the new position or the added salary regardless of the cost savings and other improvements that will result. Chapter 3 will also provide alternatives to a kaizen champion for companies that need another option. Either way, lean needs resources and people to make it happen. I will show you how to allocate the time and delegate accordingly.

Tracking

A lot of effort and time will be put into kaizen events, and it is important to track progress and the effects on the company. This fundamental aspect of project management falls on many people. I recommend that you put together a kaizen event tracking worksheet that can be used to measure event success. Remember that this tracking worksheet is to be used to measure kaizen event results, not the overall lean journey. As I mentioned in Chapter 1, it is important to encourage and allow people to improve the company all the time, outside of kaizen events. This is an event tracker only.

The kaizen event tracking worksheet has a lot of categories, and it is best to create it in Microsoft Excel. Some organizations use different software and make it available on a company intranet, visible only to employees. Either way, this tracking sheet is a live document, and everyone should have access to it to see what types of kaizen events are being conducted and how they are improving the performance of the organization. Figure 2-1 is an example of a kaizen event tracking worksheet. Customize the information in this worksheet for your company. Here are the categories I recommend:

- Kaizen event selection
- Date and length
- Kaizen event team leader
- Kaizen team members
- Preplanning
- Responsibility
- Pre-event goals
- Actual results
- Event budget
- Event spending
- Action items
- Responsibility
- Status

Kaizen Event Selection

One of the main responsibilities of the committee is to select the areas for the kaizen event. It could be assembly lines, machine work cells, maintenance, shipping/receiving, the office, or R&D, for example. There are a lot of factors to consider when selecting the area. First, look at how the process is organized and how it performs. When looking at production areas, evaluate current productivity, quality, on-time

| Kaizen Event Tracking Worksheet ||||||||
Kaizen Event	Date/ Length	Team Leader	Team Members	Preplanning	Responsibility	Strategic Purpose	
A5 Cell	Week of 3/10/08	Mark Left	Kyle Poppins	Order Wire Racks	Purchasing Mgr.	Productivity Increase	
			David Ginn	Time Studies	Kaizen Champion	Floor Space Reduction	
			Rita Pritcher	Reserve Scissor Lift	Facilities Mgr.	Scrap Cost Reduction	
			Paulina Horska	Verify Vacation Schedules	HR Mgr.	Product Throughput Reduction	
			Johnny Sherrif				
			Allen Michaels				
Maintenance	5/12/08	Gordon Black	Shawn Taylor	Service the Forklift	Kaizen Champion	Productivity Increase	
			Jose Ortiz	Buy Paint	Purchasing Mgr.	Floor Space Reduction	
			Alice Borner			Scrap Cost Reduction	
			Freya Newton			Product Throughput Reduction	

Figure 2-1 Kaizen event tracking worksheet

delivery, floor space use, and possibly travel distance. How much overtime is being worked? Do operators leave their workstations a lot? Is there excessive inventory or WIP piled up? Is the work area cluttered and unorganized (5S)?

If you select the maintenance department, the choice should be based on general organization. Are tools hidden in cabinets and toolboxes unaccounted for and disorganized? How much reactive work and "firefighting" is going on as opposed to proactive ways of working, like doing preventive maintenance? How cross-trained are the maintenance staff, and can they perform multiple jobs?

If the committee is looking to schedule a kaizen event in the office, some of the same guidelines apply. What is the level of 5S and office organization? Is the supply cabinet or room cluttered? Does paperwork such as work orders, estimates, and contracts pile up between office processes? Does the production floor constantly wait on the administrative functions to finish processing a work order or router? Often companies are looking for that golden wand that will tell them exactly where to start. You can simply look at organization, or lack thereof, and performance. Try to schedule kaizen events at least four weeks in advance.

Pre-Event Goals	Actual Results		Event Budget	Event Spending	Action Items	Responsibility	Status
Productivity	20%	28%	$500	$450	Update Work Instructions	David Ginn	Complete
Floor Space Reduction	45%	45%			Update Quality Checks	Paulina Horska	1/2 Complete
Scrap Cost Reduction	80%	95%			Finish Station Signs	Johnny Sherrif	Order Supplies
Product Throughput Reduction	45%	50%					
Man-Hour Efficiency Increase	20%		$1,700				
Floor Space Reduction	25%						

Date and Length

The traditional kaizen event is about five days long, but events can last from four hours to four weeks. It depends on the work area, goals, the product, floor space, and the level of waste. Simply place in this category of the tracking worksheet the day and week or weeks when the event will be held. It is also good to write in the hours that will be worked: 8:00 a.m.–4:00 p.m., 10:00 a.m.–6:00 p.m., and so on.

Kaizen Event Team Leader

Effectively leading a kaizen team takes experience. Any first kaizen event will be a learning experience for everyone, including the team leader. However, you have to simply pick your first team leader and go with your choice. A kaizen champion is the ideal candidate, but as I said earlier, a lot of companies do not have a champion. As a guideline, make sure the team leader is somewhat familiar with the work area. He or she should have a good understanding of waste and how to remove it. Most important, a team leader needs good project management skills and to work well with people under pressure. It is smart to develop team leader criteria to be used by the committee when making this important decision. Try also to select the kaizen team leader at least four weeks in advance.

Kaizen Team Members

Team members should be from varying disciplines and backgrounds to ensure that a good mix of ideas will be generated. Every potential team member reports to someone on the kaizen event steering committee; since the committee is responsible for selecting team members, the result should be a team with the right talents. Here are my recommendations for team members:

- Two operators or people who work in the process
- Maintenance employee
- Material handler
- Line engineer
- Quality technician
- Office personnel

- Forklift driver
- Shipping employee
- Manager

Every company is different, so base your team member criteria on your organizational chart to ensure that all the right people are on the kaizen team. Make a tentative list of team members at least four weeks before the event, and then finalize the list with about two weeks to go. This allows the company time to verify vacation schedules, those on light duty, or if the prospective team member has other scheduling conflicts.

Preplanning and Preplanning Responsibility

Traditional kaizen training typically teaches people to perform every single task of the event, from the beginning until the end. Although this makes the event action-packed, experience has demonstrated that trying to accomplish too much can have a destructive result, often placing kaizen teams in unsolvable situations. To avoid these potential pitfalls, I recommend that several preplanning activities take place four weeks prior to a kaizen event. During the preplanning activities, the target area is selected, as well as the team leader and a tentative list of members; therefore, specific planning projects can begin as well.

Preplanning involves a variety of items and activities. Contractors may need to be reserved; supplies and equipment may need to be ordered for the team; specialized tools and machinery may need to be purchased or rented; waste analysis and time and motion studies can be conducted. The number and type of preplanning activities will vary depending upon the type of event and the specific goals established for it. Solid preplanning ensures that the kaizen teams are positioned successfully.

Each kaizen event will require some level of preplanning, which is usually the responsibility of the kaizen event steering committee. A multitude of tasks need to be completed prior to the kaizen event. Special tools may need to be reserved. Employees from sister plants can be invited. Adjusting production schedules and people to accommodate the event time frame will have to be addressed. Conducting an evaluation of the work area in order to assess current state performance may be needed. During kaizen events the teams will need certain supplies such as bins, racks, shelves, tape measures, paint, and floor tape. These should be ordered ahead of time so that they are available on day one of the event. All preplanning items should be worked on at least four weeks in advance. The kaizen event steering committee members are the ones who are assigned the preplanning tasks and must dedicate time and resources to ensure that they are completed.

Pre-event Goals

During the early planning phase of any project, generating goals can be difficult. Forecasting, in any form, can be lucky or lousy. It is important that each kaizen team be faced with some moderate challenges. These events are being conducted to improve your business, so don't be afraid to set those goals! It is best practice to refer back to the established shop floor metrics, as discussed in Chapter 1, as a guide for improvements. Be sure to establish realistic goals, as unattainable goal setting will only serve to destroy the effort. An attainable goal might be improving productivity 20 percent by reducing waste in a line or a process. Yet there is no real guide for establishing your team goals. Simply set goals that you feel are realistic and attainable, and make sure that you plan adequately to ensure success.

Actual Results

After the kaizen event is complete and the workers go back to their normal jobs, the company should immediately begin to monitor progress to see how quickly the goals are achieved. This column in the tracking worksheet may not get filled in until the process or work area has consistently met the expected goals. Are the new output requirements being met? How are productivity and WIP levels? Is quality improving? Did the team save the proposed amount of floor space and reduce throughput time? It is important to realize that even if certain goals are not met, the team did not fail. As your organization becomes more experienced at kaizen and kaizen events, you will become better at estimating metric and performance improvements.

Event Budget and Event Spending

One of the fundamental elements of kaizen is that improvements should be made with little or no expense. This is true, but keep in mind that each company will have to dedicate some money for continuous improvement, and each kaizen team will need access to these funds during the event. Spending for most kaizen events ranges from $0 to $1,000, depending on what is necessary. Just allocate some money to a kaizen budget. The rate of return will be great, and any money spent will be quickly recouped from the improvements.

Action Items, Responsibility, and Status

These last three columns in the tracking worksheet are used to monitor any unfinished work from the kaizen event. Rarely does a kaizen team finish every task during the kaizen event. Minor disruptions during the project will make the team change course a little. Team members will come up with many improvement ideas along the way, and not all of them may be finished. All action items should be completed within 30 days of kaizen event completion; this is called the *30-day mandate*. The kaizen event steering committee can discuss this issue during their monthly meeting.

The kaizen champion is the person who is in control of the information in the tracking worksheet, but there are other people involved. First of all, each team leader is responsible for the success of his or her team and must report this information to the kaizen event steering committee. The forum in which everyone meets is called the kaizen monthly meeting.

The kaizen monthly meeting is scheduled, obviously, once a month to discuss the continuous improvement program and usually takes about an hour. I am not a fan of meetings, especially meetings that go long and accomplish nothing. To be honest, meetings are very anti-lean in my opinion. They take time away from value-added work and are often in place because people are not allowed to make decisions on their own. However, lean is a team effort, so it is good to meet to discuss the kaizen events and how they are affecting the company. Structure is the key here, so keep the meeting simple and to the point. To maintain simplicity in your kaizen meeting, break it down into three parts. The kaizen event tracking worksheet should be displayed during the meeting and used as a guideline for all discussions. The three parts are:

- Part 1: Previous event's results
- Part 2: Open action items from the event
- Part 3: Planning and scheduling the next kaizen event

Part 1: Previous Event's Results

Try to schedule the kaizen monthly meeting between kaizen events or just after one. As the new lean process is running, the operators will go through a learning curve. This learning curve will vary between events but it is a good time for the committee to discuss progress of the line and how the workers are adjusting. During the first part of the meeting, the committee should also evaluate how the event improved the company. Every kaizen team should have certain goals to improve metrics such as productivity, quality, throughput time, setup reduction, and floor space reduction. So at this point the committee should discuss how close the team came to achieving their goals.

The workers in the process will be the ones most affected by the improvements, as they will have to adjust to a new way of working. As waste is reduced, new ways of working have to be implemented to ensure that the "lean process" will meet their output and quality expectations. The committee needs to invite the previous team leader to the meeting and have a discussion on progress. Once the update is complete, the team leader can leave and the second phase of the meeting can begin.

Part 2: Open Action Items from the Event

Regardless of how much is completed during kaizen events, there will always be unfinished projects. Kaizen teams generate very good ideas during kaizen events to further the process at hand. Problems are presented to team members during the kaizen event, and when you put people together to create solutions, amazing things can happen. However, sometimes these ideas will take longer to implement than the time allocated for the event. These unfinished items will go on an action item list. The second part of the meeting is spent on getting a status report on these items.

All action items from an event must be completed within 30 days. Each item needs a person assigned to it and a firm deadline for completion. To ensure that the new process can produce the required results, these action items have to be completed on time. The team members who have responsibility for action items should now be invited into the meeting to discuss their progress. It is the job of the kaizen event steering committee to clear any obstacles that impede the completion of the action items and provide any support necessary. Once all action item issues have been updated, the team members can leave and the committee can begin the last part of the meeting: planning future kaizen events and other lean projects.

Part 3: Planning and Scheduling the Next Kaizen Event

The kaizen event tracking worksheet is the only document needed for the monthly meeting. During this last part of the meeting, the committee can discuss and possibly begin planning future kaizen events. As I mentioned earlier, try to schedule kaizen events at least four weeks in advance to allow for team selection and for the various preplanning items to be completed. However, it is perfectly healthy to discuss events two, three, and four months away. Just make sure, when your committee is planning future events and filling in the tracking worksheet, that if any kaizen event already scheduled is coming up on its four-week window, to start picking the teams and working on preplanning issues at that point.

Kaizen Communication

Kaizen- and lean-related activities should be continually communicated to the entire organization. As your culture slowly becomes more aware that lean is a way of thinking and working, you have to learn how to keep the momentum going. Develop what I call a kaizen communication system that works as an information delivery apparatus for your company. Often when I conduct lean assessments of organizations I find that general communication is lacking between departments. There is a clear division between the operations/manufacturing side and the administrative side. This gap in information and communication needs to be bridged to ensure that everyone knows what is going on everywhere. Your company as a whole needs to realize that it wins or loses as a team, and each department must contribute to team-based activities to improve the operation. In regard to kaizen communication, I recommend three items you can put in place:

- Kaizen communication boards
- Kaizen newsletter
- Kaizen suggestion box

Kaizen Communication Boards

I wrote about communication boards in my first book, *Kaizen Assembly: Designing, Constructing, and Managing a Lean Assembly Line*.[1] My clients have found these boards to be a very effective part of their ongoing lean communication. Simply purchase a two-sided dry-erase board on wheels. It is essentially shaped like a triangle, and it can be moved around to various places in the company.

Kaizen communication boards can be placed anywhere in the company where there is high people traffic—at employee entrances, in meeting rooms and lunchrooms, at the front entrance, or at various places throughout the factory. Depending on the size of the facility, you may have to purchase a few of them. The point is saturation of lean information so that everyone can read about upcoming projects. These communication boards contain the following information about a kaizen event:

- Area selected
- Date and length
- Kaizen team leader
- Kaizen team members
- Team goals and objectives

Kaizen Newsletter

Developing and circulating a company newsletter is a very good practice and is essential if you are embarking on a lean journey. A kaizen newsletter is devoted to the subject of continuous improvement. It can be a separate newsletter, or you can simply make space in an existing newsletter for the information needed. The key to a good newsletter is that the information relates to your specific plant. General corporate information, such as stockholder information, words from the CEO, or the acquisition of a Chinese facility, does not speak to continuous improvement. Although this information is nice, to be honest, most employees really don't care. Again, I am not trying to be negative; I just want to make the point that the newsletter should be about activities going on in your plant.

---

1. Boca Raton, FL: Taylor and Francis, 2006.

The kaizen newsletter should contain information similar to what is on the communication boards: areas, team leaders, team members, etc. More important, it should feature pictures of the team working, as well as write-ups of individual accomplishments and how their efforts are helping the plant. One unique approach to the kaizen newsletter is to allow previous kaizen team members to write about their successes. This will help further engage your people, as those who read the newsletter will read about their buddy or colleague who works with them. This is very powerful and it will speak volumes to the employees.

The kaizen newsletter should be issued monthly or bimonthly and attached to paychecks or placed in break rooms and even at the front entrance so that important visitors can read about the continuous improvement efforts going on in the plant they are about to tour.

Kaizen Suggestion Box

As a lean leader you have to encourage the production workers to tell you what needs to be improved. To help garner ideas from the floor, develop a suggestion system that allows them to provide feedback on improvements that have been made and recommend future opportunities. See Figure 2-2.

Line operators are often left out of the design and planning phase of kaizen events. I have discussed the importance of placing operators on kaizen teams, and it is my opinion that they should be involved with deciding what area is scheduled for a kaizen event. Operators are usually confined to their workstations or areas on the production floor and typically have little or no contact with management or engineers. Any contact that occurs is usually initiated by the support staff in the work area. How can you get operators and other floor employees involved

Figure 2-2 Kaizen suggestion box

in the decision-making process and get their input for continuous improvement? The employee suggestion box allows operators to give input on future improvements.

Much like a voting box, where ballots are submitted, the employee suggestion box is used to collect ideas coming from the floor, ideas that can be considered for future kaizen events. The box should be placed near the communication boards or in operator break rooms for easy access. A simple suggestion form should be placed near the box; see Figure 2-3 for an example.

When using any suggestion system, the company must develop a way to communicate back to the workers about whose ideas are being selected for a kaizen event, which ideas have been passed on to engineering, maintenance, or purchasing, for example, and which ideas will be addressed later. As this system becomes more and more popular, you will get flooded with suggestions, especially when the workers see their ideas actually being implemented. People by nature like information even if it is not what they want to see or hear. Don't leave employees in the dark. Let them know if their idea is a go or not.

Kaizen Event Suggestion Form

Employee _____ Date _____

Department _____

Improvement Idea _____

Would You Implement? Yes _____ No _____

Thank you for your suggestion.
You will be contacted as soon as this suggestion is reviewed.

Figure 2-3 Kaizen event suggestion form

Developing the individual attributes of the company kaizen program is not difficult, but it will take some time. It also has to be made to fit the culture you have to ensure that lean and kaizen will thrive. Remember the following key points:

1. Every successful kaizen program requires a firm foundation, one that allows a company to allocate the appropriate resources and make time to implement the lean initiative.
2. Establish an effective communication system that will ensure buy-in, participation, and awareness.
3. Allow every employee to have an opportunity to participate in a kaizen event and to offer input and suggestions.

Creating a foundation for change is the key. Once it is in place, go rid that waste from your operations.

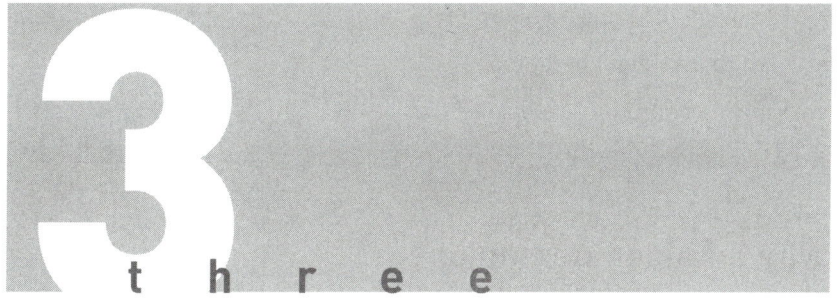

The Kaizen Champion

The word *champion,* at least in my opinion, means an individual who has the expertise to teach and guide others in the elements of a given philosophy. In sports it is basically the best of the best or the number-one team standing at the end of a season. In Chapter 2 I introduced to you the kaizen champion. This employee is dedicated 100 percent to continuous improvement and acts as the lead for the company's lean initiatives. Your organization does not necessarily have to call this person a kaizen champion. The title could be lean champion, lean engineer, continuous improvement manager—it does not matter. Before starting Kaizen Assembly, one of my titles was lean champion and the role was exactly the same as that of the kaizen champion I will describe in this chapter. I use the term *kaizen champion* in my teachings, so that is what I will use here.

The kaizen champion's role is a full-time job requiring a solid background, an understanding of lean manufacturing principles, and an ability to communicate and teach these principles to the entire organization. In this chapter I will cover the following aspects of the kaizen champion:

- Why a kaizen champion?
- Kaizen champion skill sets

- Choosing the kaizen champion
- Cost of a kaizen champion
- Kaizen champion responsibilities
- Alternatives

Why a Kaizen Champion?

My job as a consultant and author is to recommend what I think is the best approach to a successful lean journey. Although lean journeys differ from one company to the next, all of them require a significant amount of work, from training and up-front planning to implementations, follow-up, and continuous improvement. Lean journeys need fewer leaders and more doers. A good kaizen champion knows when to lead and, more important, when to get down and dirty to get the work done.

A kaizen champion is essential to a company because of the time commitment required to make lean a business model. Employees are busy with day-to-day issues in their department and are always juggling numerous projects and initiatives. Their current positions exist because their work is required to keep the company moving. Accountants and controllers are in place to handle the financial side of the business; marketing and sales people are required to keep business coming into the company and to generate new ideas for future business. Operations managers work day in and day out with production workers, production schedules, vacation approvals, delivery dates, and/or all the elements of running a production floor. I could go on and on. When day-to-day issues come first, and people are juggling numerous tasks, when do they have time to take up the full-time responsibilities of lean? Well, they don't. Now I am in no way implying that just because a kaizen champion is in place, everyone else in the organization can divorce themselves from continuous improvement. Lean is a company-wide effort. But there needs to be a guiding individual who can direct the efforts and bring in resources from other departments and functions as necessary. This guide is the kaizen champion.

With this said, I am also aware that not all companies can afford to create a new position for a kaizen champion. A good kaizen champion who can produce results is not cheap, even if the return on investment is well worth the new addition. There are alternatives, which I will discuss later in this chapter.

Kaizen Champion Skill Sets

As a business leader you always want to try to arm yourself with the best employees you can. Choosing a kaizen champion is no different, and you want to make sure you pick the right person to fill this position. I have hired my share of them and have also helped my clients to screen and select candidates. Picking one is not easy; do not rush the process. As a guide, here are the areas in which this person should have expertise:

- The seven wastes
- Lean as a business model
- 5S and the visual workplace
- Kaizen and kaizen events
- Data collection
- Setup reduction and quick changeover
- Line design and work flow
- Material replenishment
- Project management

The Seven Wastes

This one is a no-brainer, but simply knowing the seven wastes is only part of the game. Understanding where waste exists and how to reduce or remove it is another. One of the biggest mistakes with regard to waste is not knowing when it is necessary. Necessary waste is waste that is either impossible to remove or is literally part of the process. An organization has to learn to prioritize its waste reduction efforts and decide what waste is realistic to keep. For instance, workers in a company that manufactures aircraft will have to walk up and down inside the airplane to install components. This movement in and out of the plane is wasted motion, but I don't see these actions ever being eliminated. A factory will always have some level of raw material, WIP, and finished goods. These may sit around for a very short period of time, but they still constitute waste. Products will have to move from one work area to another. Automated or not, this is wasted transportation. Mistakes will be made, and even if the rework is small in nature, guess what—this is waste. A good kaizen champion teaches and emphasizes waste reduction and can show how to reduce it in an existing process. A good kaizen champion can also teach people that a totally waste-free environment does not exist.

Lean as a Business Model

This aspect of the champion is not easily learned. It is a "feeling" or instinct about how a lean company should operate. Ideally, a kaizen champion should have experience in other companies and can incorporate best practices in your company as needed. One of the fundamental elements of lean thinking is having and teaching patience. A lot of organizations get excited about incorporating lean practices and want to change everything. "Change the world in a week" becomes their way of approaching lean. It can't be done. The beauty of lean is that there is no rush. Time is completely on your side. Business leaders simply have to make the decision about when to start. My point here is that a smart kaizen champion can help bring a culture along slowly and methodically, depending on the existing level of resistance, confusion, and success. As the culture slowly develops into a continuous improvement culture, the kaizen champion can go into a "check and balance" mode rather than the mentoring and training mode that is usual at the beginning of a lean journey.

The kaizen champion must have the ability to redirect the kaizen event steering committee to ensure that any lean strategy that was created is being fulfilled. He or she knows how to garner support and resources, prepare for each project, and guide kaizen teams so that everything falls in line with the business model. Kaizen champions are like the guardian angels of lean.

5S and the Visual Workplace

5S has been brought up throughout the first couple of chapters. I am often asked how and where to start lean initiatives. It is tough to say, since every company is different in its lean endeavors. With that said, I think the implementation of 5S is always a perfect starting point for most. 5S is aggressive organization and cleanliness, a philosophy of lean that emphasizes a showroom factory. In a 5S environment, everything has a home—all tools, parts, garbage cans, tables, workbenches, pallets, carts, documentation, etc. Put this book down and think of your factory or company. Imagine a workplace where all the items have a particular location and are clearly marked and labeled. Depending on the size of the plant, it could take over a year to fully implement 5S. An experienced kaizen champion can map out a 5S implementation plan that includes the supplies and training needed and the implementation method to be used. Kaizen events are a great delivery mechanism for this, but people can start in their areas anytime.

5S should also be relatively simple for a seasoned champion, and the responsibility of training employees should eventually be shared over time. 5S essentially is the foundation needed for lean, and if a company cannot keep the house clean, then the other lean practices available will be very hard to grasp and/or maintain.

Kaizen and Kaizen Events

Lean is implemented in basically two ways. Lean initiatives can be quick or take time, depending on the goals. For instance, an inventory reduction plan can be quite intensive, involving multiple employees, departments, and suppliers. Reducing inventory in a plant takes time, and a three-day kaizen event won't do it. Line consolidation and overall plant floor reduction take time as well, especially if more product lines are being added during the same time. These are simple examples, but you probably see my point. Now kaizen events can be used to help reach certain milestones, but avoid getting into an "event-lean" approach to improvements. Don't wait for a kaizen event to reduce waste; allowing employees to practice continuous improvement is extremely important in developing a lean-thinking culture. An organization's kaizen champion can help teach people the difference between kaizen and kaizen events and will know when an event is necessary or not.

Data Collection

A multitude of data collection tools are available to evaluate any given process. The kaizen champion needs to know them and have the ability to pick the right one to conduct an analysis of current conditions, called *collecting the current state.* The current state of a process is basically the number of value-added and non-value-added steps needed to complete a product or service product. Products could be cars, microwaves, ropes, doors, lamps, stoves—any tangible goods produced. Products could also be the processed or fabricated parts, or the subassemblies needed to complete the final assembly. Service products are items such as work orders, routers, invoices, marketing plans, estimates, or anything in the administrative portion of a company. Products and service products are completed through a process. A process basically includes two types of work: value-added and non-value-added (waste). Capturing the current state involves using these data collection tools to find out what it takes to complete the product or service product from start to finish. Each step requires a certain amount of time, information that is also very important to collect. Here are three very common data collection tools:

- **Value stream mapping (VSM)**

    VSM is the visual representation of a sequence of operations and steps that occur in the process of producing a product or service product. A VSM analysis follows certain parts of a particular product family from the supplier all the way to the customer. It shows supplier delivery and delivery requirements, inventory buildup of raw material at the company, and all the individual manufacturing processes, cycle times, setup times, people, shifts, finished goods inventory, and delivery requirements to the customer. It is a high-level look at the product family from start to finish. The goal of VSM is to identify waste reduction opportunities and to establish the current state.

- **Time and motion studies**

    A good kaizen champion has the ability to conduct useful time and motion studies on work. Time studies are a very good data collection tool when manual assembly is performed or when there are machines that require setup and changeover. Time study data collection is almost an art form, as it requires experience to do correctly. It also involves interacting with production workers; a relationship must be developed first so that they trust the data collector. Don't do time studies when evaluating office work; office work has a whole different dynamic from production work. Random disruptions and inconsistent workloads from day to day make conducting time studies nearly impossible here. Time estimates are better for this environment.

- **Spaghetti diagrams**

    This tool is good for collecting the current state when workers operate multiple pieces of equipment, or if they have a large work cell in which to maneuver. Take a sheet of paper and draw the general work area. Identify the large items in the drawing: machines, shelves, tool supply areas, material storage, etc. Take this drawing to the work area in question and watch the workers perform their duties. As they walk from one place to the next, draw their paths. By the end of the analysis, you will have traced a giant bowl of spaghetti, and now you will have the total distance the workers cover in a given day. This tool is very good for preparing for a 5S kaizen event whose goal is to better organize an area for less motion and transportation. Spaghetti diagrams are good for evaluating small localized areas of a plant.

Setup Reduction and Quick Changeover

Not all processes involve manual assembly. Companies that use automated manufacturing processes where machines and equipment do the bulk of the production should have a kaizen champion who understands the importance of reducing setup and shrinking changeover times. These are tricky subjects, and it may take time to really get a firm understanding of the two. Setup and changeover are two separate steps, but they are often considered to be the same thing. Setup is the work performed before a new product run is needed. Changeover is the actual act of removing items such as fixtures and parts from the machine, and then placing new fixtures and parts for the next run. This portion of machine work must be fast and efficient so that the machine can be turned on again. The more efficient the setup steps are, the faster the changeover.

Another misconception about changeover is that it is completely non-value-added since it is downtime, but frequent changeovers are good depending on the number of different products being manufactured. For instance, if a company makes one type of phone, then changeover should be kept to a minimum. But if there are multiple types of phones with different colors and options, then the quicker a company can change over to manufacture the next model, the better it can satisfy customer expectations of delivery and quantity. A good kaizen champion knows this philosophy, can teach it, and knows how to establish smart setup steps and changeover routines.

Line Design and Work Flow

Smart physical layout of a process is also critical in keeping waste to a minimum. Reducing travel distance between operators and processes is important, as is maximizing the amount of floor space. However, making the processes so tight that people feel cramped is just as bad as giving them too much room. Your kaizen champion must first take into account the physical size of the products being built. That will dictate what type of construction is needed. For instance, if an organization produces small calculators, large, bulky conveyor rollers and belts may be a poor option. Products can be built on carts that have the appearance of a train. This arrangement makes the process very flexible, as the carts can be relocated anywhere at any time. Maybe the product is large and heavy, in which case the use of automation such as conveyor rollers or belts that are powered by a foot pedal is best. Adding a lift table to the process may help the workers get better views of the product. Proper lighting is needed. Point-of-use documentation and material are absolute requirements. The list goes on and on and I could write another book on the subject. I will go into more detail about line design and work flow in later chapters.

Material Replenishment

Material replenishment plant-wide is a complex subject, and it takes years to get an efficient system in place. It involves analyzing supplier performance with regard to cost, quality, and delivery. What quantities do you currently buy, and how often do your suppliers deliver? Are you always sold on that 10 percent discount to buy a six-month supply? But now you have to incur the holding cost of that inventory, which by the way costs much more than the 10 percent you just saved—by a long shot.

Proper material presentation in the work area is also part of this system. Obviously material should be at point of use, and the replenishment of this material should not be done by the production worker. Material and parts quantities should be kept to a minimum. Large pallets and bins of excess parts simply take up floor space and extend the length of the line. Quantities in the work area are also based on supplier performance, but customer demand and delivery frequencies also are part of the calculation. Your kaizen champion should be strong in inventory reduction and material replenishment, and he or she should have a working relationship with the purchasing department and suppliers.

Project Management

The preceding areas of expertise for the kaizen champion have been somewhat technical and analytical. Completing the champion's skill set are good project management skills. Planning, conducting, and following up on kaizen events require project management. Often a person who has good analytical skills lacks project management abilities, although this is not always the case. One of the fundamental aspects of project management is dealing with people and their different personalities. I have had the pleasure of hiring and having some of the best manufacturing and industrial engineers work for me. But if I were asked, "What was the one definable similarity among all of them?" I would say that they all struggled with working with people who had conflicting personalities. Not everything in lean is data-driven, and your people are the most important element of a successful lean journey.

Kaizen teams should be cross-functional and diverse. They have goals in front of them, and they must complete implementations on time. Opinions vary about what is the best solution for waste reduction, and sometimes kaizen events can get intense and conversations may get heated. Good project managers can deal with these circumstances and redirect as needed to ensure that people are being heard and that the team can still get the job done on time. Budgets may have to be considered as well. So much is required of a good project manager, and I feel that any strong kaizen champion possesses the attributes of one.

So there you have it. Although this was a short description of the kaizen champion, it gives you an idea of the type of individual needed for this role. Some companies take a while to find the right candidate. There is no hurry. As for any other position in the company, make sure the person you pick can fulfill the requirements of the job.

Choosing the Kaizen Champion

Now that you have a general understanding of the skill set of a good kaizen champion, it is time to begin your search. As I just said, take your time. I have a client in Chicago who took over a year to finally decide. Most of the candidates interviewed had the background and experience we were looking for, but they did not have the passion that is also needed. The company finally found someone and that person has worked out.

When selecting the kaizen champion you can go in two directions. You can select someone who currently works in the company; let us call this employee the internal option. Or you take the external option and find someone totally new to the company. Either option has it pros and cons.

Internal Option

Remember, this position is completely dedicated to lean, so if you decide to select within the organization, that person's old job must be filled.

Pros:

- In-depth knowledge of the company's operations
- Has worked with the current culture
- Is familiar with the company's products and processes

Cons:

- Is absorbed in old of ways of working
- Finds it difficult to see new approaches to working
- Is part of the old system
- Old job must be filled

External Option

If you are not able to find someone in the company to take on the role of kaizen champion, it is not a poor reflection on your current culture's abilities. The kaizen champion's job is not easy. It can be stressful at times and a lot is riding on its success. If you decide to go outside of the company, this option also has its pros and cons.

Pros:

- Brings a fresh perspective
- Has experience and knowledge from other companies
- May have a lean background

Cons:

- Is not familiar with the company's operations
- Does not know the products or processes
- Has not developed relationships with operators

Either option requires a financial investment in the future of the person and the company. If you opt for an internal person, you still have to fill his or her old position, so there is the added cost of hiring that person. Once you have a kaizen champion who can get results, however, the return on those investments will be worth the effort.

Cost of a Kaizen Champion

What is the investment? Allow me to use the external option as an example since not only have I hired kaizen champions, I was one myself some years ago. Kaizen champions who can generate results are well worth their salaries. Although the number could be more or less depending on geography and salary history, a good kaizen champion is worth around $80,000 a year plus benefits. Keep in mind that I am talking about hiring a person who is focused on one plant or at least 80 to 90 percent focused on one facility. Some organizations stretch their champions too thin if the champion has to support more than one factory. This does not mean that the champion cannot visit other plants and provide guidance, but it is best to have kaizen champions in each plant.

You are probably thinking that this approach is expensive. Well, it can be, depending on results and leadership support. But if, through the leadership of the kaizen champion, lean implementations save the company $500,000 to $1,000,000 a year from waste reduction, it seems to me that it is worth an $80,000 investment.

But don't forget the other hiring costs involved in the external option. You may need to relocate the candidate and his or her family as well. After multiple interviews, travel expenses, moving household goods and cars, lease termination fees, closing costs on selling and buying a house, and other miscellaneous moving expenses, a company could spend an additional $20,000 to $30,000 to get the ball rolling. However, with the right person, these costs will be returned 50-fold. Of course, I realize the investment must be done in the first place and it may not be economically feasible for the company; this is why the last portion of this chapter is dedicated to discussing alternatives to the kaizen champion.

Kaizen Champion Responsibilities

Once the kaizen champion is in place, you can begin to load him or her up with responsibilities. Along with the responsibilities that I have discussed so far, here are some of the other matters on which the champion takes the lead:

- Training
- Kaizen monthly meeting
- Communication boards
- Kaizen newsletter
- Kaizen suggestion box
- Kaizen event tracking worksheet
- Team leadership
- Action item follow-up
- Monitoring other lean initiatives

Training

The kaizen champion is also in charge of creating the curriculum and setting the schedule for training the company in lean and kaizen. If the company has just started its lean journey, the kaizen champion should be training all employees in lean principles. A classroom environment is the best approach, and sometimes it is a good idea to rent some space off-site to avoid disruptions. The champion needs to develop a curriculum for three demographics: First is leadership training, the second is middle- and upper-manager training (which includes office personnel and other support staff such as engineers), and the third is training at the production/maintenance/supervisor level.

As time goes on, the champion should work with the HR department to help create training for new employees regardless of their titles or where they work. This also includes creating ongoing refresher training for all employees. HR can also use the champion to develop job descriptions for new employees. The champion should always be one of the employees who helps interview and screen prospective employees. As lean becomes more and more integrated into your firm's business model, the kaizen champion should be involved in most of the hiring process to ensure that the right talent is being brought on.

Kaizen Monthly Meeting

Your kaizen event steering committee should meet once a month to discuss all lean initiatives, including kaizen events. This meeting is run by the kaizen champion, and he or she is responsible for scheduling the meeting, creating the meeting agenda, and sending out meeting notes. During the meeting the kaizen champion should get status reports from all the committee members on any lean projects in process. Discussing kaizen events is a given, as I detailed earlier in this book; even small improvement initiatives need to be discussed. It is critical to the success of your lean journey to have this meeting regardless of the amount of waste reduction effort that is going on. The kaizen champion is also responsible for keeping the other members of the steering committee engaged.

Communication Boards

I brought up the subject of communication boards in Chapter 2. It is the job of the kaizen champion to maintain this board's information and to update it as necessary to keep people informed of upcoming kaizen events.

Kaizen Newsletter

Information on the organization's kaizen newsletter comes from the kaizen champion. The champion is heavily involved in all lean initiatives and can provide valuable information for the newsletter. The champion can make the newsletter, or it can be left to someone in human resources or an office manager; it does not matter. Regardless, the kaizen newsletter is part of the job responsibilities of the kaizen champion.

Kaizen Suggestion Box

The kaizen champion should be virtually a fixed presence on the production floor and in the administrative processes, learning how they operate, developing relationships with everyone, and finding opportunities for waste reduction. The champion's presence enables people to speak openly about problems in their respective processes; basically, the kaizen champion maintains an open-door policy on improvement ideas and should be in constant discussions with floor workers about their ideas for improvements. Outside of these interactions, the kaizen suggestion box should be available for those people who do not want to speak openly. It is the responsibility of the champion to empty the box and evaluate the suggestions. These suggestions should then be brought to the kaizen monthly meeting to be discussed by the kaizen event steering committee.

Kaizen Event Tracking Worksheet

The kaizen event tracking worksheet needs to be kept up to date for all employees, and it is up to the champion to perform this task. The worksheet is presented to the committee during the kaizen monthly meetings and should be available for review by those who may not come to the meetings. Remember, this tracking worksheet is important for keeping information about all elements of kaizen events visible and accurate.

Team Leadership

In the early stages of your lean journey, the kaizen champion is the ideal person to lead kaizen teams. The kaizen champion also develops team leader criteria that can be used to select future team leaders. If you do not have a champion at a given time or at the beginning of the lean journey, simply make a list of criteria that will work until a champion is selected. Criteria or not, good team leaders are developed over time with kaizen event experiences—successful ones and not-so-successful ones. Even when the kaizen champion is not the team leader, he or she should be available to support and encourage the members of an event team.

Action Item Follow-up

After each kaizen event, there will be unfinished work that needs to be completed within 30 days. It is the responsibility of the kaizen champion to follow up with team leaders and members on their progress. The champion needs to help clear any obstacles that might be impeding completion of the items and decide if the kaizen event steering committee needs to help.

Monitoring Other Lean Initiatives

Remember, it is vital to your lean journey's success to not get stuck in "event-lean." Waiting for a kaizen event to make improvements is not the right approach. Kaizen events are one delivery option for lean implementations. What you do to improve the company outside of kaizen events is equally important. Maybe it is an ongoing inventory reduction plan or working toward a supplier certification program. Develop ongoing setup reduction and changeover teams to constantly address downtime during these procedures. Whatever the initiatives are, the kaizen champion should monitor them to ensure that everyone is able to reduce waste in his or her processes on an ongoing basis, outside of the scheduled events.

So as you can see, there are a lot of miscellaneous responsibilities for the kaizen champion. Some may appear to be simple, but I can assure you that the kaizen champion stays quite busy, and as time goes on, he or she will become highly involved in the company's overall strategic plan.

Alternatives

Some of you may be thinking, "There is no way we would be able to have a kaizen champion." Well, you are not in the minority here; and to be honest, most firms cannot. I know for a fact that a good kaizen champion will pay for him- or herself and then plenty more, but if you fall into the no-kaizen-champion category, you will be happy to know that there are alternatives. However, you must assign ownership of lean in some fashion or it will fall apart.

Think about all the requirements of the kaizen champion, and then think of those people in your company who already possess these skills. One of my clients has no champion, but the company is "doing" lean with great success. Its journey is slower than that of companies that do have a champion, but it is still progressing with waste reduction.

Assign someone to be your 5S champion. This person needs to be trained in the concepts and implementation of 5S and is responsible for making sure all projects that use the elements of 5S are successful. Your 5S champion should be involved in planning kaizen events with someone who is your kaizen event champion. This person is responsible for planning events and helping with team leadership and follow-up. Have HR take the lead on updating the communication boards and the newsletter and on reviewing the suggestions. Find someone in the company who is good at data collection, such as an industrial engineer or other technical person, who can help analyze processes and come up with better ways to flow product and calculate proper workstation and people requirements.

The talent exists in your company; just tap into it and make sure the people have the time to work on lean- and kaizen-related items. I also recommend hiring interns from the operations or engineering departments of a local college who can be used for—well, anything. A lot of my clients use interns to conduct value stream mapping and time studies while they search for a kaizen champion. Or they simply use interns on an ongoing basis for a variety of tasks. Interns are eager to use what they have learned in school and to begin fleshing out their résumés for other career opportunities. It is always good to use a consultant when needed as well (though I am not attempting to market my services to you). A consultant helps get the journey going, and a good consultant knows when to step back and let a client continue on its own.

Why a Kaizen Champion?

There is not much more I can say at this point about the kaizen champion. I dedicated an entire chapter to it because it is my honest opinion that a kaizen champion is a vital component of lean. The rate of return on this person is phenomenal. But I also live in the real world, and some organizations simply do not have this option. So I hope I have outlined some good alternatives for you so that you can decide what is best for your company. The bottom line is this: You create ownership of certain parts of lean, whether with a kaizen champion or spread out among multiple employees.

Kaizen Event Scheduling

Kaizen events require solid up-front planning to ensure that they are successful. Traditionally, it was taught that there should be no activities prior to the event. With this approach, a kaizen team would begin day one of the project with confusion and little direction. Often working 12 to 16 hours a day, the kaizen team would go through a series of trials and errors until coming up with a final layout or a handful of solutions to reduce waste. They would be exhausted and often bitter as a result of these kaizen events. I am a firm believer in and practitioner of preparation and planning for every kaizen event a company conducts. This does not mean that the answers should be provided to the team. Depending on the goals of the team, certain tasks should begin at least four weeks before the event.

Some kaizen events can be planned even further out than four weeks, depending on the complexity of the work area, the length of the line, and the amount of floor space the cell takes up; the consolidation of two processes would probably require more preplanning. As I mentioned in Chapter 2, kaizen events consist of structured time, a selected area, and a talented team. Waste removal initiatives between kaizen events are simply the practice of kaizen. These activities do not necessarily take a lot of preparation, just some effort to continue improvements. Kaizen events take a different tack.

There are basically three stages to a kaizen event: preplanning, implementation, and follow-up. This chapter will outline the fundamental aspects of kaizen event preplanning, starting at the four-week point before the event. Everything in this chapter should be used as a guideline. As your organization becomes more experienced with conducting kaizen events, you will refine the necessary steps in all stages. It is up to the kaizen committee to lead the preplanning initiatives and make sure each one is completed in time for the event.

Four Weeks Before the Kaizen Event

- Select the process/department/work area that will be the focus of the event.
- Make a tentative list of kaizen team members.
- Select the kaizen event team leader.
- Establish team goals.
- Estimate event spending.
- Order supplies.
- Update the kaizen communication system.
- Schedule outside assistance.
- Conduct waste analysis of the area.

Select the Process/Department/Work Area

Selection of the focus for a kaizen event depends on a few variables. Often "gut feel" is fine; I am an advocate of getting started and not waiting for perfect direction. Companies can be paralyzed by a lack of decision making and never truly make the move from idea formulation to action. Don't wait for the magic wand because it does not exist. However, there are a few guidelines you can follow to select areas for improvement, or at least for a kaizen event.

First, you can evaluate sales and output. Does the product or line or area contribute a large portion of your overall revenue and/or the total output? Be careful, because a high-volume product may have very little effect on total revenue as it could be a low-priced item. On the other hand, the product line could represent a large portion of revenue, but output in comparison to other products is very low. A fast-moving high-sales product is a good starting point.

Second, look at performance. Determine the current productivity of the process and compare it to other areas that may be performing more efficiently. Look at quality information internally and externally to see if the need for improvements could be based on errors, defects, rejects, rework, and customer complaints. Customer complaint information is useful as long as it is accurate. Once your product leaves the plant, it could touch dozens of hands, flow through numerous external processes, and get handled over and over again. Internal quality information is generally more immediate and has a better chance of being accurate and thus more usable. How much floor space is being taken up by the line or process due to excessive WIP and inventory? This observation or measurement will also give you some perspective on travel distance. Long distances equal longer delivery times, so it is smart to analyze floor space use and travel distance.

Third, how many of the seven deadly wastes are consuming the lives of the production workers? More important, how much of their overall day is lost to non-value-added activity? Leaving the work area, reworking products, overbuilding and overprocessing, sharing tools, and waiting on product, parts, and information are examples of things that will negatively affect cost, quality, and delivery. It will require some initial analysis of the process to get a firm grasp of how much waste there is; I will discuss this data collection exercise in Chapter 6.

Fourth, what are the production workers telling you in the suggestion system I outlined in Chapter 2? Their suggestions may bring to the surface a plethora of problems that are unknown to management and other support staff, such as problems with equipment, suppliers, parts, tools, part presentation, the order of the work, and imbalances in the process.

These four factors should be all you need to make a decision. Do not wait to have 100 percent of the information before you make a decision or you will never make one. Simply look at sales, output, performance, waste, and employee suggestions in some mix and make a move!

Make a Tentative List of Kaizen Team Members

Having all of the selected team members on the kaizen event is the sole key to success. The committee should create a tentative list of potential members four weeks before the event to ensure their availability. During this time, managers and HR can verify any vacation schedules that have been submitted for the week of the kaizen event. Also, certain employees may be on light duty due to injury or illness; this is the time to check paperwork and employee records for this information.

Another factor to consider when selecting team members is whether they are involved with other company-related activities, such as training or work-related travel. Nothing bugs me more than walking into a kaizen event thinking I have eight team members, for example, and there are only five because the proper planning was not done. So, a tentative list should be created so there is adequate time to verify that everyone can participate. Of course, family emergencies and sickness right before an event cannot be avoided.

Select the Kaizen Event Team Leader

Once the area and the team members have been selected, you can move on to selection of the team leader. Picking employees to be team leaders is not as simple as it appears. Kaizen events are very intense and require good leaders who can assess situations and direct people to ensure success. Often your kaizen champion is the best choice, but as I said earlier, not all companies can have this type of position. It is then smart to develop criteria for kaizen event team leadership. Important attributes of a kaizen event team leader are

- Project management skills
- Interpersonal skills
- Technically minded
- Good at meeting deadlines
- Ability to stay within budgets
- Effective communication of goals and objectives
- Positive attitude

When selecting the team leader, you don't necessarily have to pick the line or area supervisors. As long as the team has members who work on a regular basis in the area that is the focus of the event, your team leader can be anyone as long as he or she has the qualities listed above. Good team leaders are also developed over time, and as you circulate different people into kaizen event leadership roles, you will develop a powerful pool of change agents from which to pick for the future.

Establish Team Goals

As discussed in Chapter 2, kaizen teams should be selected and scheduled for an event with the purpose of improving the company. Without definable goals, a team of any kind can find itself working without an end. In Chapter 1, I described key shop floor metrics such as productivity, floor space, quality, travel distance, and inventory. Improvement goals with these metrics in mind will help the team identify waste removal opportunities.

Estimate Event Spending

A certain amount of money should be allocated for each kaizen event so that the team can purchase items as needed. During kaizen events, team members come up with amazing solutions to problems, and sometimes they require a quick trip to the local home improvement store. I can remember a kaizen event about five years ago when two team members devised a solution for station identification. During the event, the team was in the middle of 5S implementation, and I had tasked the team with coming up with a way to identify stations with signs. There were a couple of criteria. First, the signs had to be high enough that material handlers could see the location of the stations from a distance. Second, the station designation or number had to be visible from the same distance.

A line operator on the team and someone from maintenance got together and decided to take on the task. I was fully aware that the task at hand was not overly complicated, so I was looking for creativity. I went about my work and left to deal with another issue.

The two disappeared for a while, about two hours, and then I found them in the maintenance department. They were clearly building something for the event. At the end of the day, they came out to the assembly line with a cart full of what appeared to be white PVC pipes. They had gone to a plumbing supply store and purchased two different sizes of PVC pipes and some hardware. Figure 4-1 is a simple illustration of what they built.

Figure 4-1 Station sign

Each workstation would have one of these signs—a simple solution and very creative. They had used the money that had been budgeted for the kaizen event to buy the supplies for this idea. Usually the bulk of the event budget is for feeding the team, but there should also be enough for those small projects that come up during each event.

Order Supplies

Every kaizen team will require some supplies. Teams use a variety of things to eliminate waste or to implement a lean practice. For instance, during the implementation of 5S, kaizen teams will need floor tape, paint, labels, laminating material, markers, tape measures, and box cutters. These items are good to have for every kaizen event, and they should be placed in a "kaizen event supply box." This box is an invaluable asset to any company conducting kaizen events. It has yet to fail me or my clients. It is best to custom-build one if you have the maintenance support, but you can buy something at any home improvement store.

It is also common to order material for building things. One-inch metal tubing for special fixtures, tables, and shelves and Peg-Board for making visual tool boards are also good examples. The list could go on and on. Make sure to begin the supply-ordering phase at this point.

Update the Kaizen Communication System

When the event is four weeks away, update the communication boards, generate a new newsletter, and collect the suggestions in the kaizen event suggestion box. Start sending e-mails to the employees who are not on the team so that they know about the event. It is good for everyone to be aware of all events so that they know not to make decisions that will pull team members off the event or make the event hard to complete.

Schedule Outside Assistance

A variety of people who do not even work in the plant can be team members. Workers from sister or parent companies can be invited to contribute to the kaizen event. This is a smart approach when organizations are trying to standardize lean implementations somewhat. I say *somewhat* because every journey is different, but it still allows for best practices. Currently, Kaizen Assembly has a client with a plant in Ferndale, WA, and Lafayette, LA, and they help each other during kaizen events. (See Chapter 7.)

Other outside help could be suppliers or customers. Suppliers can be invited to attend and contribute during kaizen events that address the line for products in which their parts and materials are used. Kaizen event facilitators and consultants can be hired to lead or co-lead first-time events. All of these external resources should be scheduled at least four weeks before the event so that travel arrangements can be made.

Conduct Waste Analysis of the Area

Waste analysis will be described in much more detail in Chapter 6, but since this is the scheduling chapter, I should touch on it here. Depending on the kaizen event, data collection and waste analysis must be conducted to ensure that the team is making the process more efficient. A multitude of data collection tools are available in lean, but some are better than others for each process.

Often before a kaizen event, value stream mapping and time studies are done to capture value-added and non-value-added work. Generally, value-added work is the type of work performed to make parts, assemble products, and run machines. Non-value-added work consists of tasks that involve the seven wastes (overproduction, overprocessing, waiting, motion, transportation, defects, and adding to inventory). Value stream mapping is performed to get a high-level view of the process from start to finish, generally from the supplier to the customer. Time studies can be done to capture the work required to build a product and make a part, including the movements of people and other inefficiencies. This information can be used to pinpoint waste removal opportunities and allow the team to make major improvements to the area. Not all kaizen events require data collection, but some do. Again, I will discuss this in further detail in Chapter 6.

Waste analysis of any kind should be done at least four weeks before the event. Depending on resources, an organization should have someone focused on collecting this information on an ongoing basis to help prepare teams and to conduct other continuous improvement efforts.

Two Weeks Before the Kaizen Event

Most of the preplanning items that were started in the fourth week fold into the third week. With two weeks to go, preparations begin to change and/or speed up, depending on the task. These are the tasks that should be started two weeks before the event:

- Finalize the kaizen team members.
- Get an update on supplies and outside resources.
- Ask team members to walk through the chosen area.
- Pick a room where the team can gather.
- Analyze the collected data and start coming up with design ideas.

Finalize the Kaizen Team Members

With two weeks to go, the kaizen team should be finalized. They should know that their expertise is needed on the team and that they have been selected. Consider as you go along your lean journey that some companies get to a point in their kaizen program when kaizen event participation is voluntary. As people see the improvements being made and how kaizen is affecting the company, they will volunteer to participate in events.

Get an Update on Supplies and Outside Resources

At this point get a status report on all supplies or equipment ordered for the kaizen event. Most important, the kaizen event supply box should be stocked. Verify participation from potential outside resources who may be traveling to the kaizen event. If for some reason they cannot make it, you still have time to select others to take their place. That is not an ideal situation but it could happen, and with this approach to planning you can still have a full team.

Ask Team Members to Walk Through the Chosen Area

Have the team members spend some time in the area selected for the kaizen event. Companies that have a large manufacturing facility or multiple buildings should encourage employees to visit processes outside of their normal assigned area, especially in preparation for a kaizen event. Having the team members walk through the area ahead of time will give them a feel for how things operate. Granted, if the plant is small, employees would probably be familiar with everything already. Once the kaizen event starts, you want the team to have some idea of what kind of machines, equipment, tools, and people are used to build the products.

Pick a Room Where the Team Can Gather

During a kaizen event, the team will require two places to work. The first work area is generally the process, assembly line, or area that has been selected as the focus for the kaizen event. Sometimes working in the actual process is difficult, so pick a place where the team can work that is in close proximity to the line. The second area is a conference room or training room where the team can place their belongings. This place is also needed to conduct group discussions, eat lunch, and have meetings. Often a company has a reservation system for securing rooms. Customers, suppliers, and other employees use the rooms, so reserving one ahead of time is smart. Nothing is more annoying than not having a "war room" for the kaizen team once the week begins. Secure the room ahead of time so there is no confusion or, more important, no interruption for the team.

Analyze the Collected Data and Start Coming Up with Design Ideas

Depending on the event and the goals of the team, a current state analysis of the process may have been done ahead of time. This could have been in the form of value stream mapping (discussed in Chapter 3), time studies, waste analysis, or other flow and time evaluations. At this two-week point, start looking over the information even if it has not been completed. What can be done with this information before the event starts? Is it clear that there are too many workstations or imbalances in the work content between operators, or is there a lot of walking around? Sometimes very little can be derived from the information, but it is good to review it early. Again, I will discuss data collection extensively in Chapter 6.

One Week Before the Kaizen Event

Now it is crunch time. With one week to go, preparations had better be in the final phases.

- Gather current state information.
- Meet with the kaizen team members.
- Place all supplies in the team's gathering space.
- Meet with the plant or general manager.
- Make food preparations.

Gather Current State Information

One week from the event, get an idea of the process's current state. The kaizen event steering committee should have set goals for the team in regard to productivity increases, quality improvements, floor space reduction, travel distance reduction, setup reduction, or inventory reduction, for example. Depending on the metric(s) selected, the kaizen champion or the appointed kaizen team leader should gather this current state information, which will be presented to the team during this one-week point and during the kaizen event. The current state information provides a starting point for the team.

Meet with the Kaizen Team Members

At this point, all team members should know about their participation. If not, your company has other planning issues. There should be no confusion about who is on the team, the area selected, and the time slot for the kaizen event. Invite the team into a conference room for about 30 minutes. The team leader and/or the kaizen champion should review the following information:

- The area selected for the event: products, equipment, number of operators, output, quality and productivity issues, experience level of the operators, and number of shifts
- Team member introductions: especially smart if your organization has hundreds of employees
- Current state: productivity, quality, floor space, travel distance, inventory levels, WIP, and number of workstations
- Team goals: as selected by the committee
- Confirmation of the team's start and end times and shift

Place All Supplies in the Team's Gathering Space

This is best done the afternoon before the first kaizen event day. Place laptops, printers, the kaizen event supply box, laminators, etc., in the conference room where the team will meet on day one. When the team arrives, it is nice to have everything ready to go so they can begin work immediately.

Meet with the Plant or General Manager

The team leader and/or the kaizen champion should meet with the plant manager and give the final "go" on the project. Although the plant manager should be aware of any changes at this point in the game, it is always healthy to meet with this individual and go over any final thoughts. The plant manager needs to provide total support for the project, so any final words of wisdom from either the team leader or the plant manager should be said at this time.

Make Food Preparations

One of the most influential elements in getting employees to volunteer for events is the food provided by the company during the event. Assign someone to take care of ordering, delivering, and setting up lunch or dinner for the team. Don't be cheap with this, and be creative. I have a client in Chicago that brings in a barbecue during summer kaizen events; steaks and hamburgers are cooked on the shipping dock for the team. One time the company brought in an ice cream cooler packed with ice cream bars and popsicles. Pizza and deli sandwiches are probably good enough, but make sure to provide a variety of food for the team.

As your organization has more and more kaizen events, and more people become team members, everyone will get this fringe benefit. Try to avoid going out for lunch, because this tends to be a time-consuming affair requiring transportation and eating time. The lunch is also a good time to have your midday meeting anyway.

Final Thoughts on the Timeline

The recommendations I provided in this chapter are simply guidelines. Depending on your culture, management, and kaizen event success, your organization can develop its own timeline and preparation structure. Each of the items I described is listed because some delay during the 200 or so kaizen events in which I have been involved occurred because one of these tasks had not been done. Kaizen event leadership is really about project management, and any good project manager will say that good preparation never fails. Once all the preparations are complete, it's "go" time!

Part II
Kaizen Events

The next chapters will use everything I have previously outlined to explain the different types of kaizen events. Each chapter will be specific to the "theme" of the event. The proper planning criteria will be addressed in each chapter; however, keep in mind that preplanning items will vary depending on the theme of the event. I will still try to keep some consistency with what was discussed in the kaizen event timeline portion of Chapter 3. Each kaizen event chapter will follow the five-day event format. Just remember, not all events are five days long. Length will vary depending on the amount of work to be done. Chapter 5, 5S Kaizen Events, will describe the planning and execution of a kaizen event used to implement 5S. I will give examples of a production process and a maintenance department. Chapter 6 describes a standard work kaizen event, which involves more than the implementation of 5S. I will cover how to conduct preliminary evaluations on the process like time studies, line balancing, and inventory replenishment.

Chapter 7 provides a real-world example of the concepts in this book. It is a case study of Samson Rope Technologies, which successfully implemented a company kaizen program.

5S Kaizen Events

As mentioned earlier, kaizen events are one of the mechanisms used to implement lean. In essence, each event needs a "theme" with clear goals and objectives. The purpose of a 5S kaizen event is to implement 5S in an area in need of cleanliness and organization. This chapter is designed to show you step by step how to conduct a 5S kaizen event.

5S implementation is always a good starting point for a newly established lean journey. Companies can be as aggressive as they want, but implementing the 5S philosophy is easier than other lean practices. Because the results of 5S implementation are visible, it provides the "tangible" element of lean, easier to see than, for example, a new changeover procedure or total preventive maintenance.

Using the kaizen event approach is equally beneficial because it brings together a cross-functional team of individuals who together can implement 5S in a way that is effective and highly visible. In this chapter and the next, I will provide timelines for preparation, and then for each day of the event as well. There will be variations in the planning portion of each chapter. You may want to refer back to Chapter 4, which discusses preparation for a kaizen event.

Four Weeks to Go

Preparing for a 5S kaizen event is probably less difficult and time-consuming than preparing for other types of kaizen events. It does not really require much data collection such as value stream mapping or time studies. Sometimes it is good to perform a spaghetti diagram analysis prior to the event, but it is not required.

Select the Area

Selection of the area depends on a multitude of variables. Often the need to open up more floor space is a prime reason for doing a 5S implementation, especially when a new product or new product lines are coming and the company wants to avoid constructing a new building. With that said, I think it is important to quickly mention the concept of cost avoidance. Often organizations get caught up in cost reduction. Although reducing cost is a positive side effect of lean, the benefits of cost avoidance can be tremendous. Through an aggressive 5S implementation, the company may be able to avoid adding onto an existing building to accommodate a new product line and the added inventory, and so avoid spending a huge amount of money. In addition to the cost of the construction, there are the taxes and the ongoing monthly expenses associated with a new facility. So, with that said, sometimes the reason for conducting a 5S event in a given area is to open up enough floor space for the growth of the company.

The spaghetti diagram analysis could reveal a significant amount of wasted motion and transportation. This analysis could provide insight into how to rearrange the work area for more effective movement. The choice of the area for the 5S kaizen event can simply come down to the fact that 5S has not yet been put in place. You really can start anywhere. Maintenance, R&D, shipping, the office, an assembly line, a work cell with machines—it does not matter. The length of time required for a 5S kaizen event depends on the amount of floor space the work area consumes and the amount of "stuff" on that floor space.

Select the Team Leader

As I mentioned in the previous chapter, the kaizen champion is the perfect candidate for team leadership, but anyone with knowledge of 5S is fine for leading a 5S kaizen event. You can pick the area supervisor, an engineer, or a lead operator; whoever it is must be trained in 5S and ideally has seen it implemented somewhere.

Tentatively Select the Team Members

As for any kaizen event, a cross-functional team of employees is needed to help foster new ideas and thinking "out of the box." Always have people on the team who work in the chosen area, especially production workers. You can also include administrators, engineers, supervisors, line workers, maintenance personnel, and forklift drivers. The kaizen event steering committee, which is responsible for getting the right mix of individuals, will pick this team during the kaizen monthly meeting. This list is temporary until team member participation can be verified in regard to vacation time, light duty, family responsibility (depending on the shift worked), and other activities happening in the building.

Establish Goals

Each kaizen event is designed to improve the work area and the company and to reduce waste. 5S is a very powerful improvement tool that has the ability to reduce all of the seven wastes, as well as improve productivity, reduce floor space and product travel distance, and enhance quality. It is sometimes hard to quantify the results after a 5S implementation. I have no doubt that there will be major metric gains after 5S if the team does its job correctly. Goals for a 5S kaizen event could be floor space reduction, travel distance reduction, and productivity gains. The most common is floor space reduction.

Event Spending and Supplies

To get the 5S program started, a kaizen event supply box needs to be constructed and supplies such as floor tape, paint, label makers, stencils, tape measures, etc., purchased. So the start-up costs for your first event will be higher than for subsequent events. Most of the money invested in a 5S kaizen event is for stocking up on 5S supplies and maybe miscellaneous building material such as one-inch metal tubing, Peg-Boards, and other items.

Update the Kaizen Communication System

Once the event has been scheduled, the champion or whoever is responsible must update communication boards and the kaizen newsletter so that everyone is aware of the event.

Identify the Kaizen Team Meeting Space

A 5S team will need a place close to the work area selected where the laminator, some laptops, a printer, and the team members' personal items can be put. Ninety percent of the team's work will be done in the work area that is the focus of the event, but the team will also need a place for making large labels, laminating, and storing personal items such as water bottles, food, etc.

Schedule Outside Assistance

If your organization has multiple plants and the first 5S kaizen event is the first for the entire corporation, then invite employees from sister plants to witness and participate in the proceedings. They can learn valuable information from the event which they can begin to apply at their respective facilities. This approach is smart if you are attempting to standardize the 5S program throughout multiple factories. It is also good to bring in salespeople, suppliers, and customers at some point during your kaizen events.

Two Weeks to Go

Finalize Kaizen Team Members

Of course, this is a guideline for all kaizen events; a 5S event is no different. Verify that the participation of all team members who have been selected is confirmed. If there have been some changes, make the appropriate substitutions now.

Get an Update on Supplies and Outside Resources

With two weeks to go, the 5S supplies that will be used during the implementation should also be finalized. Make sure that all necessary supplies are on their way or have arrived. Certain supplies will be purchased during the event, but it is smart to check on everything ordered at this point. It is also good to confirm the participation of those who do not work in the plant who are coming in for the event, especially if they are included in the final roster of team members. Airline tickets, hotels, and car rentals should be arranged at this point.

Ask Team Members to Walk Through the Selected Area

Ask the final 5S team to start spending time in the work area selected. If they have been fully trained in 5S and all of its implementation techniques, they will see ample opportunity for improvement. This walk-through may not be needed if the facility is small. But some companies are quite large, and it is always smart to allow the team to spend time there as the event gets closer.

One Week to Go

Gather Current State Information

Most of the more complex preplanning tasks should be nearing completion by the one-week point. With a week to go, the kaizen champion should collect some vital information about the work area selected for the 5S implementation. Most kaizen events will require some current state information in regard to floor space use, product travel distance, productivity, quality, and inventory levels. Once improvements are made and the workers have worked within a new 5S environment, these key shop floor metrics should be positively affected. Have the current state information available for the team on day one of the kaizen event.

Meet with the Kaizen Team Members

This meeting is scheduled by the event's team leader and/or the kaizen champion. It is a formal meeting to allow team members to meet each other if they do not already know each other. It is also an opportunity for the team leader to discuss the goals for the team as outlined by the kaizen event steering committee and go over any other relevant information regarding the project.

The team leader can discuss what types of preparations were made to ensure that the team will have the time, resources, and supplies needed to meet their goals and objectives. Team members should be allowed to ask questions and provide feedback on any observations they have made prior to the kaizen event.

Place All Supplies in the Team's Meeting Room

This particular task should really be done the day before the event begins. If the kaizen event is to start on a Monday, the team leader and kaizen champion should place all appropriate supplies, equipment, the kaizen event supply box, and other miscellaneous tools in the team's chosen work area on the preceding Friday. Printers, laminators, label makers, and even laptops can also be placed and made ready the day before the event begins.

Meet with the Plant or General Manager

It also good for the team leader and the plant manager to meet just before the event. The team leader confirms that the kaizen event is a "go," and the leader and the plant manager can then discuss any final preparations. This is not the time to tell the plant manager that the event is to be canceled for whatever reason. The need for a cancellation should be apparent long before this meeting, but if not, that would mean that the preplanning fell apart in the four- and two-week timeline. Events are usually canceled because preplanning items did not get completed.

Let It Begin!

The ultimate goal of any 5S team is to have the chosen area "5S-compliant" by the end of the kaizen event. Essentially this means that every item that is required to perform the work in the work area has a home, and the location and item are clearly marked. It is important to complete each S in the implementation in order before moving on to the next—or at least to do the best you can.

Day One: Sort

The team leader should break the team into two sub-teams to begin the Sort portion of 5S. During the sorting phase, all items deemed unnecessary should be removed from the work area, as well as any items that are questionable. Questionable items are items that are not used very often, as opposed to everyday necessities. The team needs to decide what the home location will be for infrequently used items. The two sub-teams are as follows:

- Sorting team
- Collection team

Sorting Team

The sorting team is responsible for sifting through the workstations to identify all unnecessary tools, supplies, tables, chairs, garbage cans, etc. It is best to use the production workers who were picked to participate on the sorting team. Their detailed knowledge of the workstations can help the other team members. Pair up a production worker with someone who does not work in the area. This allows for an outside eye to play "devil's advocate" and question the items in the workstations.

To conduct the sorting activity, the team should use what is called a *red-tag campaign*. A red-tag campaign is an organized approach to sorting; it allows a lot of people to be involved in the sorting process and it keeps items being taken from the workstations organized. There are three parts to a successful red-tag campaign:

- Red tags
- Red-tag area
- Red-tag removal procedure

Red tags are visual indicators that a kaizen team has "tagged" the item and deemed it unnecessary to perform the work in the workstation. It is literally a red-colored tag. The sorting team places these tags on items, removes the items, and brings them to the red-tag area for further evaluation. Figure 5-1 shows what a red tag might look like.

During kaizen events, quick decisions must be made at the sorting phase because the bulk of the work is done during the second phase, Set in Order. This can be an emotional time for the team members, as people become very attached to their work belongings. During a five-day event, the red tagging/sorting needs to be completed by the end of day one to ensure that the area is 5S-compliant at the end of the week. Anything can be removed during a red-tag campaign. Possible items are:

- Extra air tools
- Extra hand tools (wrenches, screwdrivers, socket sets, etc.)
- Workbenches
- Lights
- Fixtures
- Garbage cans
- Chairs
- Documentation
- Cabinets and shelving
- Parts and material

Figure 5-1 Red tag

The list could go and on and on. Cabinets, drawers, and tool chests should be completely emptied out, and items should be sorted to identify only the bare essentials. Always start with the small items. And I am serious about sorting things like extra pens, pencils, and wrenches. As insane as it may appear, a collection of these small items requires medium-sized storage such as bins, organizers, shelves, and tables. Medium-sized items require large storage such as workbenches, cabinets, tool chests, and racking. Large items require floor space, and finally floor space requires buildings and facilities. Do you see my point?

As mentioned before, one of the main goals of a 5S team is to open up floor space, to use current floor space better to reduce the waste of motion and transportation. From a growth perspective, opening up existing floor space is essential for adding new product lines and products to avoid adding onto the facility and incurring more cost.

Place red tags on all items pulled from the workstations, fill out the appropriate information, and place the items in the red-tag area.

Collection Team

The collection team is required to be in the red-tag area, which is a temporary staging area for the sorted items. At this point in the process, nothing has left the building; it is too early for that to happen. The collection team receives everything coming into the red-tag area and organizes things based on the information on the red tag.

Red-Tag Area

The red-tag area should be marked off with red tape and a sign hung to identify it clearly. The collection team should inventory the items brought there so that the team as a whole can decide on the fate of each item. The inventory list provides insight into the amount of money tied up in unnecessary supplies, tools, workbenches, etc.

Most of the first day will be spent sorting, and the team should focus on its completion by day's end. During lunch, the team leader needs to get an update on progress from each sub-team and provide support, including shifting people between teams if necessary. One of the critical attributes of a good kaizen team leader is the ability to evaluate progress and make sure the team members are working on the correct items at the right time.

Ideally, by the end of the day one, the team should be in a position to begin thinking of the next *S* in the 5S implementation. The second *S* is Set in Order, sometimes called Straighten. One of the best preparations for Set in Order is to completely clear out the work area so that the actual floor space is empty, making it easier to see the available space. This allows the team to piece together the work area as they want so as to use the floor space more effectively. This is a grand opportunity for the employees who work in the area and are on the team to set up the area as they see fit. The team leader ensures that reduction in motion, transportation, and product travel distance and better use of floor space are being considered. The best situation after day one is to have the workstations and work area completely sorted, the red-tag area organized, and the work area completely cleared in preparation for the beginning of day two, Set in Order.

Red-Tag Removal Procedure

After the first day, the team leader and/or kaizen champion should begin putting together a removal procedure for the items in the red-tag area. This task does not need to be completed during the event, but when creating this procedure there are a couple of criteria that need to be established:

- Deadline

    How long do you want to hold on to your junk? Do not get into a vicious cycle of moving garbage from one place to another in the factory. Establish a deadline for removal: 30, 45, 60 days—something. I have seen some red-tag areas last one week. It depends on the items and what kind of "bond" the company has with the items. If anything is now deemed unwanted, get rid of it.

- Removal options
    - Auction the items off to employees or simply give them away.
    - Have a "garage sale."
    - Donate to local nonprofit organizations and colleges.
    - Give items to a local recycling organization.
    - Send the items to a sister plant.
    - As a last resort, throw them in the garbage.

Rid yourself of the stuff and move ahead!

Day Two and Day Three: Set in Order and Scrub

As mentioned before, make sure all the sorting activities are complete so the team can see what remains to be organized. The Set in Order phase takes the longest and often can be the most tedious of the work. The goals of Set in Order are to organize the area so that everything has a home and to improve the flow of incoming and outgoing material/products/finished goods. Most important is the need to reduce floor space use and product travel distance. It really does not matter if you are working on an assembly line, a work cell with machines and equipment, or a shipping department—you can apply Set in Order to all of these.

Always go into the second S with the mind-set of making everything visual and accessible. Opening up floor space and work surfaces is the key to any 5S implementation. I always tell my clients to "go vertical" and use dead space. Dead space is everywhere. Companies think that flat surfaces are needed to store and hold supplies and tools. Often, workbenches and tables are brought into a work area to hold things. But almost anything can go vertical on tool and supply boards with a little creativity. I recommend that you avoid using cabinets, drawers, and shelving as much as possible. I also realize that some things do need to be in a cabinet with a door, like delicate calibration devices, but the problem with doors and drawers is that they hide things, and unnecessary items will accumulate very quickly.

Before you can get into the fine detail of Set in Order, begin with the floor items. Decide what surfaces are needed in the workstations and area to be used for the actual work. Always evaluate the size truly needed to perform the job at hand. The team should come up with a few layout designs to choose from. Once they decide what the flow of material and parts will be, they can begin piecing together the process. Begin with the large items:

- Work surfaces: workbenches, conveyors, tables, etc.
- Garbage cans: identify the minimum needed but also take into account point of use to reduce motion
- Locations for staging of incoming and outgoing material, parts, and products
- Miscellaneous items: vacuums, welding machines, and specialty equipment and machines

Any items that will go on the floor should be the first items placed in the work area. Nothing really is permanent at this point, so don't start marking the floor with designations and identifications; try to make some tentative decisions about placement.

This portion of Set in Order may take part of or even all of day two. Once the team has decided on the locations of the floor items, they should lay out all the necessary tools, supplies, and various workstation-related items in the workstation to get a look at what needs to be organized. By going into the Set in Order phase with the mind-set of going vertical and avoiding flat surfaces, you challenge the team to come up with creative and innovative approaches to organizing. Only as an absolute last option should the team bring in a cabinet, shelf, or tool chest. The key to 5S is visibility, and tool chests and cabinets do not allow you to see where things are and what is missing.

Break out the kaizen event supply box, Peg-Board, paint, cleaning supplies, and any miscellaneous supplies. The Set in Order and Scrub phases can be done together. Always paint when possible to create a showroom appearance. Cleaning can only get you so far; painting workbenches, shelves, metal stands, etc., can make the area appear very clean.

Once the floor items are placed, it is time to create designations or "addresses" for the floor items that will hold supplies, tools, parts, etc. Using floor tape, make an outline around the items on the floor to identify their location. The best approach is to mark off those items that can be moved. Items that are bolted to the floor or heavy workbenches that will not move do not get floor tape, but everything else should be taped. Once the taping is complete, create the designations/addresses. Figure 5-2 shows a top view of what a work area would look like with floor identification.

As you can see, there is a clear outline of the items on the floor. The kaizen team should place labels or stencils on the floor describing the address. In this example, A1 and A2 are used. A1 and A2 are locations for important work-related items that are needed in the workstation. Whatever goes there, that will be its location. So if a calculator, tool, or bin full of material goes in A1, the item is also physically marked "A1."

If these items are not returned to their locations for some reason, anyone who notices that they are out of place can quickly find their home. This approach to "addressing" is extremely powerful in saving time and reducing motion and confusion, and it can help reduce purchasing and replacing tools and supplies. It is also important to place the designations up high so that they can be seen from a distance and not just on the floor.

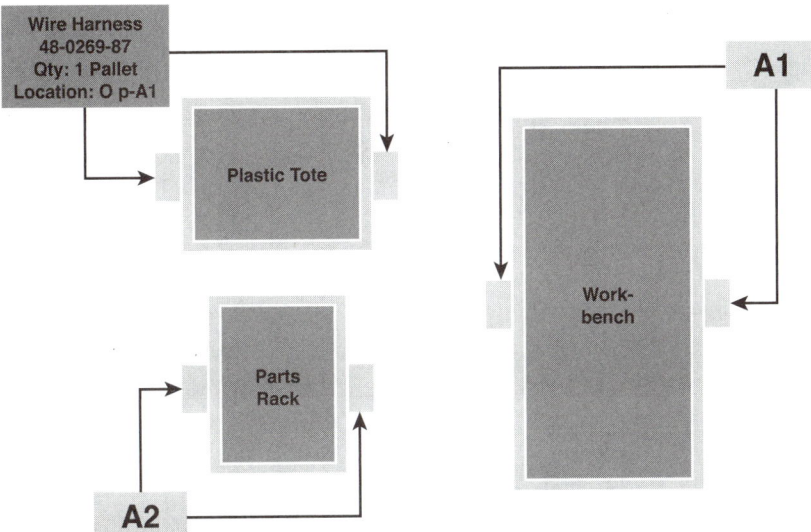

Figure 5-2 Floor identification

Tool Boards

Tools should be placed vertically on tool boards, visible and accessible. Often called *shadow boards*, these boards keep tools off the work surfaces and allow the operators to see them. You can get very creative with these tool boards. Figure 5-3 shows what one should look like.

Here are some tips for making a tool board.

1. **Paint the board:** Use any color; I have seen red, yellow, black, blue, green, and orange shadow boards. The team leader should pick some team members to begin painting the Peg-Boards at the beginning of day two so that they have time to dry before being used.

2. **Lay out the tools:** Once the boards are dry, lay them on the floor or another flat surface and lay out the tools and supplies on them. And I mean everything: tools, tape, scissors, calculators, clipboards, whatever. Your supply box should have pegs, double-back tape, Velcro tape, and other things that can be used to hold items vertically. You are basically making a cookie-cutter tool board to identify the exact surface needed to hold the workstation's supplies.

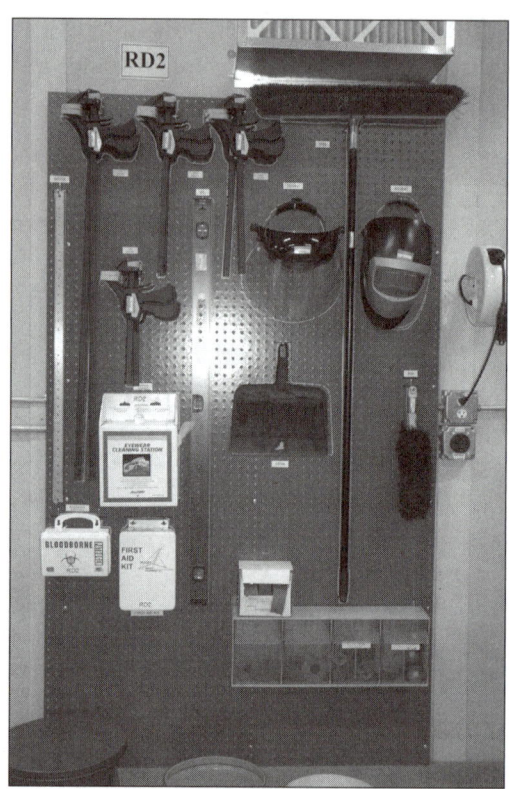

Figure 5-3 Tool board

3. Cut the Peg-Board: Now remove the tools from the Peg-Board and cut a nice square piece for the workstation.

4. Install the tool board: If a tool board will be hung on a wall, use 1-by-1 studs or some other kind of wood frame to hold it. The pegs that will be used to hang items need clearance behind the board to lock into place. Tool boards can go anywhere. I have seen them installed on the side of workbenches and cabinets (if present), and on casters (wheels). As long as the location is accessible, at point of use, and does not disrupt the operator, the placement is good.

5. Hang the items: Now the team members can hang and organize the tools and supplies on the tool board. Leave room for labels between the items and enough space to draw a shadow or outline of each one. This process is relatively slow but it is critical to any 5S implementation. After placing the tools on the board, use the paint pens that should be in the event supply box to outline each tool on the

board. When the tool is off the board because it is being used or is missing, these outlines will make the tool's absence visible. Next, make a label with the name of the tool and place it near the tool's location on the board.

6. The last step is to designate a location for the tools, such as B5, J7, L3. This "address" must physically go on the tool as well as the label so that people know where it belongs. Figure 5-3 is a good example of a tool board that was built using these six simple guidelines.

Parts, Material, and Supplies

The kaizen team must continue organizing all remaining items. Always thinki about reducing special needs. Use square bins to organize items that cannot go vertically on boards. When placing anything on a flat surface, mark off the area, generally using smaller-width colored tape. All items must be marked with their home location, and a label must be placed on the flat surface identifying what goes there. Everything—everything—has a home.

In a traditional five-day kaizen event, the Set in Order/Scrub portion will take about two days, give or take a couple of hours. There may also be some miscellaneous sorting that goes on. The team will begin to see the value of reducing horizontal surfaces and using minimal space and will question everything as they go along.

Day Four: Standardize

Day 4 is for fine-tuning and ensuring that the team is completing all the tasks in a way that is consistent. For instance, all tools should be hung vertically on Peg-Boards, the backs should be painted, there should be an outline of every tool, and the tools and board should be labeled appropriately. The tools on each board may be different but the general appearance is the same. Also, all items on the floor should be clearly marked with a location and identified with their names. The uses of floor tape should be consistent. Maybe yellow is for caution, black is for finished goods, and red is for items in the workstation. Try to standardize your approach.

The team leader should begin to create the report-out presentation that will be given to the company to outline the team's accomplishments.

The presentation consists of the following items and will be presented on day five:

- Picture of the team
- "Before" pictures of the area
- List of team members' names and titles
- Goals and objectives: reduction of floor space, product travel distance, motion, etc.
- "After" pictures
- Individual team member accomplishments
- Lessons learned
- Action item list: unfinished items to be completed in 30 days

Day Five

Try to schedule the report-out presentation for a time when a lot of people can attend. On day five the team leader completes the presentation and can take any final pictures of the area to be used in the report out. Team members can use this time to clean up the area, finalize any labeling or other 5S items, and put together an action item list of anything that did not get done.

This action item list, or 30-day mandate, outlines what each unfinished task is, who is responsible for completing it, and the deadline for completion. Companies that struggle to see things through to completion can use this mandate as an opportunity to teach their people to better follow through on tasks. Completing the tasks on this list is vital to the success of the team and those who will now be working in a new area.

After the presentation to the company, the team should invite the attendees to the work area that now has a new look. A tour helps people to see the tangible element of lean and allows them to ask questions and give comments based on what they see. Explain the importance of this organization and how it benefits everyone. Congratulate the team on their accomplishments and get some rest.

Maintenance 5S Events

Maintenance departments are perfect candidates for 5S implementations and practices. Often these departments employ multiple people doing various repairs, preventive maintenance, and special projects. Their work area is a community space where tools and supplies are shared. Sharing tools and supplies in an assembly line where there are definable workstations is very dangerous. It promotes lots of wasted motion, the chance of losing tools, and lost production and concentration. Work in these types of manufacturing processes is clearly outlined, and the workstations should have exactly what is needed to perform the work.

Maintenance departments do not operate under these guidelines. Although there should be defined schedules for regularly performed preventive maintenance, there is still plenty of inconsistent work and special projects. My point is that there is often a need for duplicates of tools or supplies when they are shared. This type of environment is ripe for 5S.

The implementation of 5S through a scheduled kaizen event is similar to what would be done in an assembly line or other part of the production process. 5S can be implemented outside of kaizen events; the implementation is slower, but the outcome is the same. Let me describe a 5S kaizen event for a maintenance department.

As for any other event, make sure you follow the preparation guidelines that were outlined at the beginning of this chapter: the four-week, two-week, and one-week requirements. The only real difference is that I usually advise my clients to sort a little earlier, about a week before the event. Maintenance commonly becomes the dumping ground for things over time, and maintenance personnel like to stock up for what-ifs. Often these what-ifs never materialize, and unused items just build up.

In a typical five-day event, most teams should be given the first day for sorting, but I recommend that a maintenance department start this activity early, being fully aware that the kaizen team will do more on day one of the event.

Day One: Sort

The team leader should break up the team into two sub-teams, as described earlier in this chapter. One team places red tags on those items being removed, and the other team places the items in the red-tag area. The creation of the inventory list for this type of 5S event is critical because a lot of expensive items may be eliminated. Final removal of items from this red-tag area may take longer; it may take more time just to decide what to do. Don't be afraid to remove that $10,000 brake press if it is not used anymore. This can be very hard for people because of its monetary worth, but if it is not used, remove it. The space it occupied can be used for equipment that is frequently used or just to make more work space.

Maintenance departments need space to work on projects, so the existing use of space needs to be challenged. I have seen 5S kaizen events in maintenance that were so successful that two separate departments were consolidated into one. Sort, sort, sort!

Day Two and Day Three: Set in Order and Scrub

Let the insanity begin. Break out the paint on this one for sure. The kaizen team job here is also to create showroom conditions, and often cleaning the maintenance department is not enough. There are often a lot of stands, equipment, workbenches, and other items that are made of steel tubing and metal. Paint away, and I recommend bright colors like yellows, reds, and light blues. I remember a kaizen event where all the team did was sort and paint. Without rearranging the area at all, they incurred a 10 percent increase in productivity and output. Bright, well-lit work areas just breed better performance. Go into any fitness center and you will see my point.

Maintenance departments love their tool chests and cabinets. As in any 5S event, the goal is visibility, so I guarantee there will be some soul-searching here. Empty out all tool chests and cabinets after sorting and get an idea of the number of tool boards that will be needed and the items that can simply be hung. Those cabinets and tool chests hide things, collect junk, and are conducive to plenty of extra sifting and searching.

All tools must be visible and hung vertically to save space. However, if there are delicate testing and measuring devices that do need to be behind doors for protection, then so be it. Everything else must be visible. Start constructing the tool boards as described previously in this chapter.

While some team members are putting together tool shadow boards, other team members can be cleaning and painting equipment and reorganizing shelving that may be needed to hold big and bulky items such as saws, drills, and heaters—larger items that cannot be hung vertically. The key here is no doors.

When organizing items that sit on shelves or multiple shelves, have the team make cubbyhole sectional pieces to maximize shelf space. Often in shelving there is a lot of dead space between the shelves. Maximize this space by make sub-shelving, and you just might see a tremendous amount of consolidation. Sorting includes substituting unnecessarily large tables and storage areas with sizes that are smarter for their intended use.

As in most five-day events, the Scrub and Set in Order phases will take at least two days and may even last into the fourth day.

Day Four: Keep Going

That is the best way to put it. Keep labeling, painting, marking the floor, and creating home locations. The team may want to use spray paint and stencils to mark off floor locations, as many maintenance departments can be dirty environments. There may be a lot of welding, dust, oil, debris, and hazardous material, so laminated labels on the floor may not work.

Day Five

Spend the last day of the event completing any unfinished labeling and start cleaning up. As mentioned before, conduct a report-out presentation and tour of the area.

5S Sustaining Tips

My last recommendations in this short explanation of a 5S event for maintenance are guidelines that you can incorporate into the department to improve performance. Just like maintaining anything new in a culture, sustaining the 5S program is hard. Your sustaining efforts will never end, including continually improving upon what was already implemented, but each company must find a way to do it. Here are a few recommendations:

- Create an end-of-day cleanup procedure.
- Conduct a daily/shift walk-through.
- Establish a 5S audit sheet.
- Create and maintain a 5S tracking sheet.
- Develop a 5S incentive program.

Create an End-of-Day Cleanup Procedure

For each area, put together a list of tasks workers must complete roughly 15 minutes prior to leaving. This cleanup procedure goes beyond simply sweeping the floor and dumping garbage. Some possible items that could be added to the procedure are the following:

- Empty all garbage and recycling bins.
- Sweep the work area.
- Return tools to their designated locations.
- Return supplies to their designated locations.
- Place pallet jacks, garbage cans, chairs, and hand trucks in their designated locations.

I recommend that you post these procedures and allow the operators time to conduct the cleanup to help sustain the improvements made.

Conduct a Daily/Shift Walk-through

Each area supervisor should take a few minutes after everyone has left to walk through the work area and verify that the end-of-day cleanup was completed and all items were returned to their home locations. If your company implements 5S to the detailed level described in this chapter, this supervisor walk-through should be quick. Any small deviations from 5S can be quickly resolved during that walk-through.

Establish a 5S Audit Sheet

If you are still looking for more ideas or a better way to sustain 5S, then you can incorporate a weekly or monthly 5S audit. Develop a 5S audit sheet with information similar to the cleanup procedure. Figure 5-4 is an example of a potential 5S audit sheet that can be used.

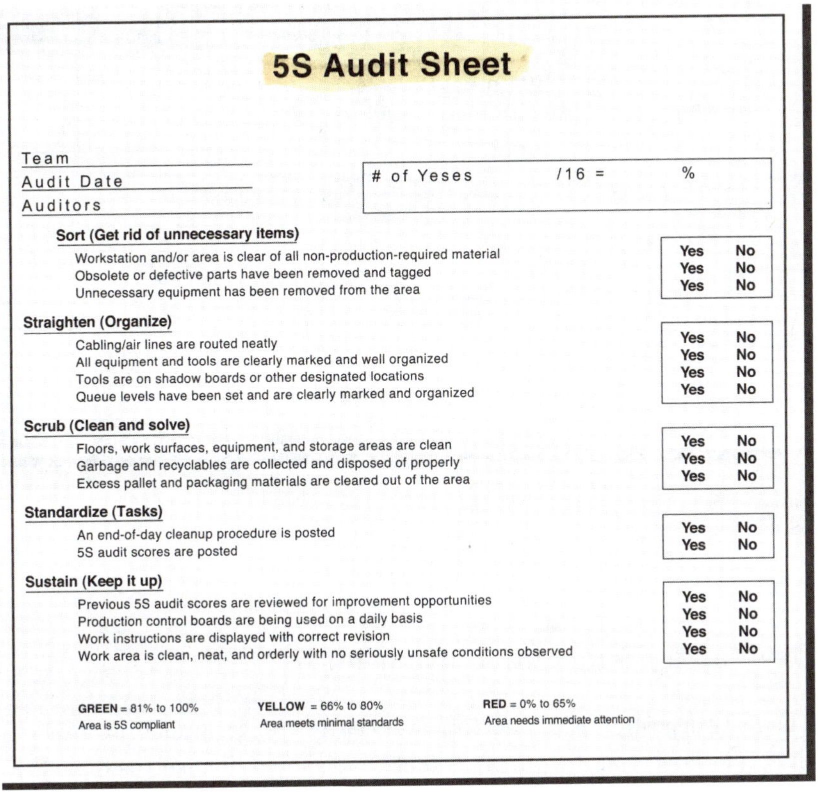

5S Audit Sheet

Team _____
Audit Date _____
Auditors _____

# of Yeses _____ /16 = _____ %

Sort (Get rid of unnecessary items)
- Workstation and/or area is clear of all non-production-required material — Yes / No
- Obsolete or defective parts have been removed and tagged — Yes / No
- Unnecessary equipment has been removed from the area — Yes / No

Straighten (Organize)
- Cabling/air lines are routed neatly — Yes / No
- All equipment and tools are clearly marked and well organized — Yes / No
- Tools are on shadow boards or other designated locations — Yes / No
- Queue levels have been set and are clearly marked and organized — Yes / No

Scrub (Clean and solve)
- Floors, work surfaces, equipment, and storage areas are clean — Yes / No
- Garbage and recyclables are collected and disposed of properly — Yes / No
- Excess pallet and packaging materials are cleared out of the area — Yes / No

Standardize (Tasks)
- An end-of-day cleanup procedure is posted — Yes / No
- 5S audit scores are posted — Yes / No

Sustain (Keep it up)
- Previous 5S audit scores are reviewed for improvement opportunities — Yes / No
- Production control boards are being used on a daily basis — Yes / No
- Work instructions are displayed with correct revision — Yes / No
- Work area is clean, neat, and orderly with no seriously unsafe conditions observed — Yes / No

GREEN = 81% to 100%   YELLOW = 66% to 80%   RED = 0% to 65%
Area is 5S compliant   Area meets minimal standards   Area needs immediate attention

Figure 5-4 5S audit sheet

Create and Maintain a 5S Tracking Sheet

Based on the scores from the 5S audits, display the results on a tracking sheet that is visible to the whole company. This creates awareness and healthy competition between the areas, and everyone can see how the plant is doing overall. Figure 5-5 is a simple example of a 5S tracking sheet. It becomes a nice visual reminder of the progress being made with 5S.

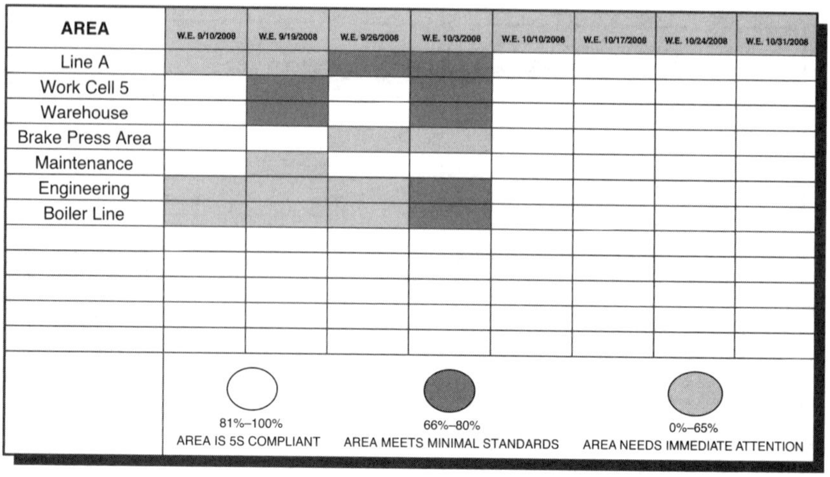

AREA	W.E. 9/10/2008	W.E. 9/19/2008	W.E. 9/26/2008	W.E. 10/3/2008	W.E. 10/10/2008	W.E. 10/17/2008	W.E. 10/24/2008	W.E. 10/31/2008
Line A	░		▓					
Work Cell 5								
Warehouse		▓		▓				
Brake Press Area								
Maintenance		░						
Engineering	░			▓				
Boiler Line								

○ 81%–100% AREA IS 5S COMPLIANT ● 66%–80% AREA MEETS MINIMAL STANDARDS ◐ 0%–65% AREA NEEDS IMMEDIATE ATTENTION

Figure 5-5
5S tracking sheet

Develop a 5S Incentive Program

The last suggestion is to develop an incentive program that rewards those employees and work areas that have shown themselves to be the champions of 5S. Hand out quarterly incentives like gift cards, pizza parties, or bonuses to the area with the highest 5S scores for the quarter. Areas not receiving an incentive will catch on fast and begin to make more of an effort to sustain their areas and continually improve on what has already been put in place.

Standard Work Kaizen Event

A *standard work event* is an industry name for the transformation of an old line into a more continuously flowing process that incorporates single-piece flow, 5S, new work standards, and time standards. Preparation for this type of kaizen event is quite different from that for a 5S kaizen event. When preparing for a 5S event alone, the kaizen steering committee must simply pick the area and load up on supplies to be placed in the supply box. Not much more is really needed. Standard work events require solid up-front planning, usually including calculation of the following:

- Effective time
- Volume requirements
- Takt time
- Process analysis

Preplanning

Effective Time

Effective time is often called *touch time*; it is the amount of clock time in the day that is available for operators to perform their work on the line. Here is an example.

In the building	480 minutes
Start-up	−10 minutes
Morning break	−15 minutes
Lunch break	−30 minutes
Afternoon break	−15 minutes
Cleanup	−15 minutes
	395 minutes, or 6 hours, 35 minutes effective time

Any new assembly line or process should be designed to achieve required volume and output based on this amount of time. Effective time should be the true measurement of productivity and performance.

Volume Requirements

This number needs to be thought through. Using historical data, firm orders, and forecasting, the company must determine the required output that can be accomplished in a full day. That is the number that will be used to design the new line. When volume requirements fall below that design rate at any given time during the year, manpower and workload changes are simply made and fewer people are needed to operate the line.

Takt Time

Takt is a German word meaning "rhythm." *Takt time* is the pace of the line required to achieve the design rate based on customer requirements. The calculation of takt time is the following:

$$\frac{\text{Effective time (shifts worked)}}{\text{Output requirement}} = \text{Takt time}$$

$$\frac{395 \text{ minutes (1 shift)}}{20 \text{ units}} = 19 \text{ minutes, 45 seconds}$$

Basically, if a line is required to build 20 units a day at full capacity, then one unit must be completed every 19 minutes and 45 seconds in the effective time available. The output requirement can be any unit of measurement (pounds, feet, pallets, etc.).

Process Analysis

Many different process analysis tools are available. In this chapter, since I will be describing a traditional manual assembly line, time studies and line balancing will be outlined.

Time and Motion Studies

Before any design or construction can take place, you must first select the assembly line and product that will be put through the rigors of a kaizen event. Second, the work associated with building this product needs to be timed. Time and motion studies are an absolute requirement for improving the assembly operations in the factory and should be done about three weeks prior to the event. I am a firm believer in making decisions based on data; good solid data can never fail you.

Time is everything. Time and motion studies are the foundation of any manufacturing process improvements and have been an early stumbling block for many companies. Top management generally perceives time studies as a wasted task. I don't agree. Time studies should be at the top of your preplanning list, but they can be difficult. Many facilities either lack this information, or the time studies were acquired incorrectly.

Traditional kaizen practice is to conduct the time and motion study on the first day of the kaizen event. Generally, this data is collected very quickly. There are a select few who can conduct time studies effectively. Ideally the kaizen champion or an industrial engineer should be assigned to the task. It takes time to master the art of collecting data, and time studies are no different. However, to help avoid difficulties, you should follow some simple rules.

1. Use a stopwatch.
2. Use a time study collection sheet.

3. Document the tasks and elements before timing.
4. Capture every piece of work content, both value-added and non-value-added, from start to finish.
5. Involve the operators.
6. Conduct the time studies.
7. Capture at least eight samples of every task or element.
8. Time all product options.
9. Remove the highest and lowest times and calculate an average.

Use a Stopwatch

A time studier obviously needs a stopwatch to perform the work. Use a standard digital stopwatch that can be found at any sporting goods store. Find one that will last and be reliable. It is best to use two stopwatches at the same time. As the time studier becomes more proficient at the work, two stopwatches will make the task more efficient so that more than one element of the work can be timed simultaneously.

Use a Time Study Collection Sheet

Raw data should be documented and saved for future reference. You will see that a time study collection sheet is very easy to use. It contains all the information needed for designing an assembly line. Microsoft Excel works well for setting up such a sheet, which should contain the following columns:

- Sequence number
- Work content
- Value-added work
- Non-value-added work
- Sample quantity
- Average

Document the Work Content

A mistake that many people make when performing time studies is trying to capture too much information at once. It is important to remember that collecting this data takes time, and you should never rush the process. I have learned from experience to write down the work content first and then return later to time. Once you are confident that the work content, both value-added work and non-value-added work, has been identified, simply go back and time all the steps.

Capture Every Piece of Work Content

Documenting work content is not as easy as it may appear. There is usually a lot of value-added and non-value-added work associated with installing even one part. I recommend that the work content be documented from start to finish through the whole process until the last workstation is analyzed. Do not get caught up with stations and number of operators. Create a long sequential list of all the work required to build the product. The line will be drastically changed and new station requirements will be established with this information.

Involve the Operators

Seek the advice of the operators during the data collection phase. Let them know that you need to know what work is performed at their station. Allow them to discuss problems that hinder their ability to perform efficiently, and ask lots of questions about their responsibilities. Take a general interest in what they do, and let them know that you are collecting data to help them in the future.

This is also a good time to let the operators know that you will be conducting time studies on the work content. Explain that it is not an attempt to see how fast they are, but to document how long it takes to build the product. If you present yourself professionally, taking into account their concerns, when you return to time them, you will get their support.

Conduct the Time Studies

Now that all the work content is documented, return to the first workstation and start timing. You should have spoken with all the operators at every station. Be smart in your timing. Do not hinder the work of the operators, because they still have to work while you are timing them. Some people are uncomfortable being timed because they feel they have to work faster. Use your best judgment when timing; if an operator is clearly stumbling about or working too fast, go to another station and come back later. Since you have all the work content written down, you can really time anywhere within the assembly line. Reassure the operators that this is not about speed and that their name and station will not be on the time study collection sheet. This is why I previously advised not getting caught up in operator names and stations. The operator will feel more comfortable knowing it is an anonymous task. Just make sure the operator is experienced enough to provide an efficient time for designing an assembly line.

Capture at Least Eight Samples

Timing work content once or twice will not yield accurate data for design purposes. Operators are faced with obstacles and challenges all day, so their work does not always take the same amount of time to perform. Time the same work content eight times to ensure that different situations are accounted for. Always stop timing when the operator drops a tool, looks confused, or walks away to talk with someone. Try to distinguish the abnormalities that would not be considered during assembly line design. This does not mean to skip timing non-value-added work like walking and waiting. Capture all the work involved in building the product and separate the useless acts.

Time All Product Options (if Possible)

Many companies offer their customers various options for their products. Some options are used so often that they are virtually standard; however, some options are very rare. It is important to document and time all options regardless of their frequency. I realize that this may be difficult depending on what is being built at the time of your observations, but do your best. There may come a day on the new assembly line when there is a large number of a rare option, and if it was not accounted for in the design, there will be bottlenecks, workstation imbalances, quality problems, and possible line shutdowns. Time the worst-case scenario for the product.

Remove the High and Low

After all eight samples have been timed, remove the highest and lowest times, unless the high and low times are close to the remaining six samples. They represent the rare and unusual circumstance. Do not confuse the high and low times with times for infrequent options. An example of an unusual circumstance that would result in an unusually high time is if an operator is struggling to fit a part into the unit because of a one-time defect from the supplier. Although such things will happen from time to time, do not include them in the design. Take the remaining six samples and average the times. This average will represent the standard time for the work.

Time studies are so important, and the accuracy of this data will make or break the new assembly line. Errors in time studies will show up very quickly after the new line is up and running. Do not rush this analysis, but then again do not take four months to conduct the study. The time study collection sheet will be the foundation for designing an effective assembly line, so take care with the information that will go in it.

Line Balancing

Once the time and motion studies are complete, look over the information and come up with waste reduction ideas. Try to discover possible ways to reduce the amount of time people leave the workstation to retrieve parts and tools. Identify potential rework reduction opportunities and better ways to present tools in the workstation. Make a list for the kaizen team to review.

About a week before the event, try to generate options for preliminary line balancing. When balancing workloads among stations, there are a few rules you can follow that can make the process easier:

1. Balance by time
2. Balance by work content
3. Balance by inventory

Balance by Time

The first pass of balancing is to use the time study information to add up the time associated with the work until takt time is reached. More than likely, the times will add up to a little less or more than takt time because of the nature of the individual time standards. Don't worry; this is just the first pass at balancing. What you are basically doing is identifying potential workstation work content. Continue with this exercise until you have gone through the whole sheet.

Balance by Work Content

The second phase of line balancing is to identify the work in the time study sheet that can be shifted from one workstation to another without affecting quality. Basically, you are moving the old sequence around to help balance the workloads in the new line. This step takes a little more thought and requires operator and engineering involvement. Is there work that used to be done in station 1 that can be implemented in station 3? By moving work around, you can better balance the line as a whole.

Balance by Inventory

Your last pass at balancing is to buffer the stations with inventory. For instance, if building the product requires a testing procedure that is double the cycle time of takt time, then make space in the workstation to test two units at the same time. As long as product is heading into the testing area and product is leaving every takt time, the process will still be balanced.

The kaizen team can look over the line balancing information and come up with further improvement ideas; however, conducting the exercise of balancing at least will give the team and the company an idea of workstation requirements to support volume.

Day One

The team leader should start the kickoff meeting by introducing the team members to one another. If the company is relatively small, the team members may recognize each other. Each team member should explain what his or her title is and where he or she works in the facility. The team leader should spend about an hour discussing 5S, standard work, the seven wastes, and visual management. This orientation will help the team members get a better understanding of why kaizen is important and what tools and methods they will be using to meet the objectives of the event. Even if people have been through training sessions in lean methods prior to the event, it is still good for the team leader to review the tools so that everyone has a fresh perspective on how to proceed.

Make sure the team understands any constraints that may exist related to moving machines and changing the plant layout. The team will have to be smart in their approach so that operators are allowed to continue working. Good teamwork will make the event more enjoyable as well as productive.

After introductions, the team leader should begin the kaizen event overview and describe the objectives the team will be required to complete by the end of the week. The team leader should already have completed a schedule for the week, which should be posted on a flip chart or a presentation board. The events of each day should be outlined with a list of the intended objectives.

Try to stagger the start time so that the team is working during the last few hours before the operators leave. They can interact with each other about issues before the line workers go home and the kaizen team starts moving things around. The schedule is a general outline because activities during the event may change.

All team members should be present when the event starts, during meetings, and at the end-of-day closeout meeting. Since team members were selected two weeks prior to the event, there should be no excuses for absenteeism. The first day of the kaizen event involves two major activities. First, sort out all the unnecessary items in the old line by conducting a red-tag event. Second, review the line balancing information, or standard work sheet, that was prepared the week before, and search for errors and other improvement opportunities. Split the team into two sub-teams to work on the two activities. The team leader should assign roles to each team member as well as their respective goals. Both teams should reconvene at a designated time for lunch to give a brief status report.

Red-Tag Team

5S should be the cornerstone of any kaizen event regardless of its theme. Here is review of the 5S portion.

Sorting does not necessarily mean just throwing things away. In a future event, items removed during the red-tag campaign might find a use somewhere else. Sorting is a method of identifying potentially unnecessary items in the facility, assessing the need for them, and dealing with them effectively. Select an area in the plant where tools, workbenches, jigs, dies, rejects, parts, and any other items that are no longer used can be placed. This will be the red-tag area. Tape off the area with red tape and post a sign indicating its use. This area is off-limits to everyone in the plant except for the kaizen team.

The team leader should hand out a box of red tags that the red-tag team will use to identify the items to be removed. I have seen many successful kaizen events where the sorting activity was conducted without red tags. However, once the kaizen event was complete, it was difficult to evaluate the items and decide what needed to be done with them. A red-tag event simply organizes the sorting activity in a more productive manner.

During the red-tag campaign, team members will be approached by operators who will challenge the removal of an item. Listen to their concerns. If the operator is unable to validate the use of the item, a red tag should be placed on it. Since every kaizen team includes an operator from the assembly area, identifying red-tag items should be easier and less stressful.

The removal of WIP during the first day can cause a conflict. Although WIP is non-value-added, allow the operators to work through the WIP during the kaizen event. Day two and day three of the week are good times to pull out WIP from the line and filter it back through during the implementation of single-piece flow.

The sorting team should spend the first half of the day placing red tags on unnecessary items. They should report back at lunch with information on their activities and results.

Review the Line Balancing Information

Once the sorting team has been released to the production floor, the team leader should hand out the line balancing information to the remaining team members, who will be the review team. Again, having operators from the line on the team will be a very valuable resource for sifting through the information. They may find errors that can be resolved prior to assembly line construction. The review team should look for opportunities for waste reduction and mistake-proofing. Make sure to have the original time and motion study data. The kaizen champion should have removed a lot of the walking data with the assumption that the parts and subassemblies will be placed at point of use. Also, the packaging removal operations will now be done in the receiving area, not by the operator, so that work content will not be included in the design.

Part of the review process is to go out to the assembly line and verify the work content and time studies. This is also a good time for some people on the review team to go to the floor and document the non-value-added work. They can also use the old time and motion data to find areas of waste. This exercise is good for verifying that the line balancing information is accurate. The review team should take stopwatches, randomly select a few operators, and time their work. This exercise will be the final check of the data, and any changes to the line balancing sheets should be done during this time. The review team should spend the first half of the day looking over the standard work sheets and analyzing the assembly line.

The team leader should leave the two teams alone and allow them to work on their assignments. This is a good time for the team leader to take current state photos of the assembly line for the report-out presentation. The team leader should also participate in both activities to ensure that the work is being done effectively.

Completing the Red-Tag Event

Workbenches and large equipment will need to be removed. Maintenance personnel on the team should make sure that equipment is properly disconnected and that hanging wires and cords are safely secured.

The team needs to move quickly and start pulling the red-tag items away before operators remove the tags. People become accustomed to working in waste because it creates a buffer for their inefficiencies.

A red-tag campaign will provide insight into the amount of waste that has accumulated over time. The red-tag area will fill up quickly. Two team members should be assigned to the area to monitor items as they come in and verify that the red tags have been filled out properly. These two people can help organize and maintain order so that items do not simply pile up.

During the red-tag removal phase, the members of the kaizen steering committee and other top management should be invited to see the mountain of waste that has accumulated. The team leader should take photos of the red-tag area for the report-out presentation on Friday.

Eliminating unnecessary items through the red-tag campaign will open up a tremendous amount of space where future assembly lines can be installed. As red-tagged items disappear from the assembly line, little islands of workbenches and equipment will start appearing. They represent the maximum number of items needed for the work to be done. By the end of the first day, the kaizen team should have successfully completed the red-tag event and finalized the line balancing information. The operators will be a little confused the next day. Just remember that each event is thoroughly planned four weeks in advance, and the production supervisors and operators are aware that there will be some added stress during the event.

End-of-Day Meeting

The day should conclude with an overview of the day's results and setting of the action items for the second day. The team leaders should leave information for the production manager outlining what has been done to his or her work area.

Day Two

The team should start the day as always in the breakout room at the time designated by the team leader. Day two is the first day of laying out the new line. The team will conduct the second and third Ss of the 5S program: Scrub and Set in Order. Maintenance projects should also begin on day two. The first assignment of the second day is to go over the line balancing information and come up with at least two line design ideas. Depending on how big and complex the line is, sometimes it is good to have these line design ideas done before the event. I have even conducted line design kaizen events to make sure that the implementation team had everything they needed for construction. However, most of the time there is plenty of time during the implementation event to come up with line designs that will work. Just use your best judgment when making this decision.

Line Design

Depending on floor space restrictions, the kaizen team should draw up at least two line designs. U-shaped cells are very good for assembling small and simple products. Operators are usually positioned in the middle of the cell, working together to build a product. U-shaped cells allow the operators to see and help each other. The cell acts as one autonomous team. The first and last operators stand back to back.

U-shaped cell operators are generally more flexible and can shift from station to station.

However, not all assembly "lines" can be converted into U-shaped configurations. Product lines that are big and bulky may require physically bigger assembly workstations and material storage, which would not allow operators to have close contact with each other. Also, U-shaped work cells should not consist of more than eight workstations. I have seen U-shaped cells that were quite large because of the size and quantity of parts in the cell. The operators were not close and could not see one another well.

It is a good idea for the team to draw up the line design on CAD software. Make sure that they draw in any expansion joints and retaining poles that are on the production floor. If your company does not have a plant layout, measure the area where the old line used to be, and stay within those boundaries. The red-tag campaign from the previous day should have freed up a lot of floor space, so it should not be difficult to fit the new line layout into the area.

After discussing the preferred design, the team leader should assign the action items for the day. The assembly process needs to be pieced together into the shape of the new layout. The team leader should hand out one or two standard work sheets per team member, and the team members should go out onto the floor and measure the items needed to construct the line—part racks, workbenches or lift tables, computer stands, pallets, bins, totes, garbage bins, etc. Again, the standard work sheet will outline everything that is necessary.

The team leader should assign a team member to act as a coordinator and monitor the measuring activities. As information comes back from the team members, the team's drafter can start drawing the line in the CAD software. The goal should be to have the measurements and line layout completed prior to lunch.

Midday Meeting

After lunch, the team should meet and discuss the action items for the second part of the day. Putting together the new line will take two days. Five stations a day is a good goal. There are two objectives for the team. First, the team will construct the workstations. Second, the workstations will be set up with the required parts and tools.

Scrub/Shine

Before the team begins construction of the assembly line, clean the floor to give it a showroom shine. Many factory floors are painted, and I highly recommend this for the sake of appearances. Painting a factory floor can be very expensive and should be done at some point in the life the factory, but do not attempt this during a kaizen event because the paint will not cure in time. Computer equipment, tools, fixtures, machinery, and any other mechanical objects should be cleaned as well, not only for appearance but for functionality. Workbenches, shelves, tables, and storage racks need to be clean and free of dust and debris. This act of cleaning is the Scrub/Shine function of the 5S campaign.

Maintenance and Machine Shop Projects

The team needs to decide, using the standard work sheets, whether there are any projects that the maintenance department should begin. Any new or existing equipment should be checked to be sure that it operates properly and is ready for installation.

The assembly line should be pieced together starting from the last workstation. Find a starting point to measure for correct placement of the line; expansion joints or retaining poles work well. Allow four feet of space from the workstation to any material storage. This will allow the assembler to easily turn and maneuver within the workstation. Let the material dictate how big the workstation area should be. Do not design every workstation to the same dimensions. Some workstations could have large bulky parts, and others could have small brackets and hardware.

Tool Presentation

At this point the area will still be a little cluttered. Excess parts and tools left over from the old assembly line will be lying around, taking up unnecessary space on the new assembly line. Although the red-tag campaign will have eliminated a majority of the tools, more than likely part quantities may still be too high. Split the team into two groups. One group should work on tool presentation and the other on part presentation. Tools and parts should be presented in similar ways.

Ideally, tools are positioned over the operator's head and at arm's length. This is good for small handheld air tools. Large tools will need more complex positioning. Non-air-powered tools, such as socket sets, wrenches, wire cutters, etc., should also be at arm's length but necessarily overhead. Have team members take the standard work sheets out to the new workstations and formulate ideas for tool positioning. The operators who are included on the kaizen team are the best resource for this exercise. There are a variety of options for tool presentation, and every workstation will be different depending on the tool requirements. The two most common approaches to hanging tools overhead are tool balancers and tool retractors.

Tool balancers are good for positioning hand tools overhead. Hand tools are connected to the balancers with small clips, and the operator can simply pull the tool down to do the work. Once the operator has finished the work, the tool balancer will simply pull the tool back into position overhead. However, there is one negative aspect to tool balancers: They tend to tug on the tool, and sometimes the operator feels as if he or she is fighting with it. It is good practice to use lightweight hand tools when dealing with tool balancers.

Tool retractors are a great alternative to balancers. They are very similar in regard to placement and function; however, they are designed to lock into position once the tool is pulled down. This enables the operator to maneuver around the product without fighting with the retractor. The operator then pulls down a little to unlock the retractor cord and the tool can roll back into the static position.

Shadow Boards

Small tools and other devices needed for the assembly line may not be good candidates for overhead presentation. Shadow boards like those described in the previous chapter can be used to store tools in an organized fashion. They can be installed on the parts racks next to the operator. A missing tool becomes very noticeable at the end of the day.

Subassemblies

Space needs to be allocated for finished subassemblies within the main assembly line. As with parts, quantities need to be established for subassemblies. Selecting the correct bin and container sizes is critical to keeping floor and shelf space to a minimum. A variety of sizes and shapes will be used, depending on the part.

Miscellaneous items such as pens, operator identification, labels, etc., should also be assigned a bin. After the bins have been selected and the team has filled the parts to the required quantities, a parts rack needs to be chosen. Mobile parts racks of some kind should be used when possible. A variety of racks can be ordered from material storage suppliers.

Day two is action-packed for the kaizen team. The team should meet in the breakout room before leaving to discuss the results of the day. The team leaders should explain to the team that day three will go more smoothly, because all the training on workstation construction and part and tool presentation happened on day two. The team should be able to complete construction of the assembly line on the third day.

Day Three

As always, the team should gather in the breakout room and discuss the action items for the day. Since all visual aids, signals, labels, and other visual management tools will be done on day four, the operators need to look around to see if they have questions about how the workstations are set up. Even though the line is more organized than before, without the correct visual management it may become disorganized very quickly. Have some of the team members on the line until the shift is done for the day. The team needs to apply the same dimensional rules as they did on day two to complete the line.

The team should take its usual lunch break and have a midday meeting. At the end of the third day, the line should be ready for the implementation of visual management. The team should gather in the breakout room and discuss the day's events. Did they complete line construction? Has maintenance finished installing the tools and creating the shadow boards for each station? The team leader should take the team out to the production floor and walk through the line. Although there is one more day of work to go, the team should be happy about their accomplishments.

Day Four

Visual management is the key to running a lean assembly line. It reduces the amount of firefighting and provides real-time information on the progress of the line. The key to visual management is to be creative. We use visual aids every day and sometimes forget that they are even there. Imagine what it would be like driving to work every day without road signs, speed limits, road dividers, turn lanes, car signals, off-ramp signs, and all the visual aids we use on the road. A manufacturing floor should be set up to run itself and allow management and engineers to react to the visual aids. Visual management is generally associated only with making production information visually available to track progress, but it goes further that that. Not only does it provide real-time information on daily performance; visual management is about having the appropriate signals and cues to allow the manufacturing process to operate on its own, with people directing the processes as needed.

The final day of implementation will be busy. However, it is a day of creativity and thinking out of the box. The visual controls and systems will be implemented in the assembly line, allowing it to operate virtually unattended. The team leader should go over the action items for the day and determine who is responsible for completing them:

- Creating workstation and parts rack signs
- Floor taping and designations
- Determining subassembly build levels
- Installing shadow boards and tower lights

Creating Workstation and Parts Rack Signs

This is a simple but somewhat time-consuming task. Each workstation and parts rack will need identification. One team member should be responsible for making and installing all workstation and parts rack signs. An accurate count of all workstations and parts racks needs to be done first to make sure no workstation or parts rack designations are duplicated. It is good practice to identify workstation numbers in the assembly line. I have seen hundreds of workstations over the years that were labeled based on the work content performed in them. In a way, this implies that the work in the workstation will never be rebalanced. Workstation numbers are best. Print them on 8½ by 11-inch sheets of paper and laminate them for protection. The signs serve two purposes. First, they identify the workstation, and second, they are parts location designations. Try to install the workstation signs high enough that they can be seen from a moderate distance.

This exercise is easy and the team leaders should allow the team members to come up with creative ways to identify the workstations and all parts racks on the floor. As long as there are no duplications, anything really is acceptable.

Floor Taping and Designations

This exercise will require two team members. Anything that is sitting on the floor has to be identified in the same manner. Lift tables, parts racks, pallets, disposal bins, workbenches, the exit conveyor, etc.—all should be identified. Floor tape comes in a variety of colors and styles. Yellow floor tape is the most popular as it creates a bright outline around the items on the floor. After every item is outlined with yellow tape, the two team members should create floor signs that identify the items. All labels, signs, and other identification should be laminated to protect them from damage. Floor signs may become dirty or torn. The lamination material will protect the signs, allowing them to last longer.

Make sure the surfaces are clean, and use double-sided tape to secure the signs to the floor. It is also good to place clear packaging tape over the floor signs for added protection.

Floor signs for parts require a part description, part number, and quantity. Parts racks and workstations should be similarly designated. It is good to place two floor signs for all the items on the floor so that they can be seen from any side. This exercise should take the two team members most of the day to complete.

Determining Subassembly Build Levels

If the proper visual aids are not put in place, the operators may overbuild items that are not needed and slow the main assembly line down. In conditions like this, it is good to create build levels for the operators. This tells them when to stop and start building subassemblies. They can switch back and forth as a team, building the appropriate quantity at the appropriate time.

The team leaders should select people to create and implement the visual aids for the subassembly operators and for the main assembly line operators who install them. There are two designated racks in the subassembly work cell for placing finished goods. Based on the quantities above, build level signs should be made to direct the operators.

Installing Shadow Boards and Tower Lights

Maintenance should have the shadow boards ready for installation by day four. Since most of the air tools are hanging above the workstations on retractors, the shadow boards will be small. They can be installed on the parts racks of the workstations or any other area that is effective.

Tower lights are a critical aspect of visual management. They are very common but are frequently used incorrectly or not at all. Tower lights are the communication system between operators and the rest of the plant. There is a variety of tower lights on the market, and every company attaches different meanings to each color.

Red can be used for signaling that there is a major problem in the workstation. There could be a quality problem or a tool or lift table malfunction; the operator may have run out of parts or needs to speak to the production supervisor. When the red light is on, any support staff in the area must drop what he or she is doing and address the operator's concerns.

Yellow can be used when there is a minor issue. This color can also be used to signal a material handler that more parts are needed in one bin or a subassembly is getting low. The material handler can go to the workstation in question and find out if the operator needs material assistance.

A green light indicates to the entire facility that everything is operating fine and the unit will move within the desired takt time. Implementing visual management is very important in a lean environment. Without the required direction and information on daily progress, management and engineers will slowly return to reacting to problems as they occur. Instead, support staff can react to the signals that are outlining the deviation from the standard, and problems can be resolved before they happen.

The team should try its best to complete the items listed in this chapter on day four. About an hour before the end of the day, the team should finish in the breakout room. The team should take a moment to reflect on what they have just accomplished. The team leader should make a list of the individual achievements of the team to include in the report-out presentation for day five.

Day Five

Closing out the kaizen event is an important part of any project. This is the team's opportunity to showcase their accomplishments. There are two parts to a report-out presentation: the presentation itself and a tour.

The report-out presentation should be scheduled in the late morning so that the team leader has time to prepare. The purpose of this presentation is to show the entire facility the accomplishments of the team and the improvements that were made to the business metrics.

The team should be allowed to get some rest and come to work as needed. The team leader may need to meet with the kaizen champion (unless the kaizen champion led the event) to go over how the presentation should be put together.

The report-out presentation also includes a tour of the new assembly line. This will give the people in attendance a chance to see the line operating under its new conditions. The presentation itself should take around 30 minutes, and the walk-through should last until all questions or concerns have been answered. The presentation should include

1. Names of the team leader and team members, and everyone's title
2. Name of the assembly line or area
3. Kaizen event date
4. Anticipated results
5. Actual results
6. Before and after pictures
7. Lessons learned
8. Thirty-day mandate (to-do list)

The 30-day mandate is a list of items that were left unfinished from the kaizen event. Every event will have a to-do list, and team members should be assigned to complete the items on the list within 30 days. There are rare occasions when these items will require more than 30 days; however, the kaizen steering committee should try to assist the team in completing their tasks within the time frame.

seven

Case Study: Samson Rope Technologies, Inc.

When I wrote my first book, *Kaizen Assembly: Designing, Constructing, and Managing a Lean Assembly Line*, I dedicated one chapter to a case study of a company that used my company kaizen program as the foundation and driver for all lean initiatives. The company was highly successful in its journey, and today I use the same model for other clients. Of course every journey is different, and implementations are always customized to provide the greatest possibility of success for each organization.

The company kaizen program illustrated in this book has not changed a whole lot; I have made just a few modifications over the last three years. The company described in *Kaizen Assembly* was a manual-assembly-based manufacturer with no equipment or automation. This chapter will be dedicated to describing the lean journey of Samson Rope Technologies, Inc., headquartered in Ferndale, WA, with another facility in Lafayette, LA. Samson Rope is not the traditional manual-assembly-based operation. It operates highly automated processes to manufacture commercial-strength rope for customers in the commercial marine, offshore rigging, commercial fishing, arborist, utility, safety, rescue, and recreational marine industries. Samson has been around for more than 100 years and is recognized as the leader in developing and manufacturing high-performance ropes. The company's unwavering commitment to research and development and its unique package of field engineering and after-sale support services have resulted in stronger and more durable products for a diverse range of commercial and recreational users. Samson manufactures over 1,000 different products, many of which come in a wide range of configurations, such as twisted, plaited, and braided, and they vary by diameter, color, length, and terminations.

Here is the story of how they used our company kaizen program approach. The names and content of this chapter were approved by the organization prior to publication.

January 2007

It was winter 2007, and I had just completed relocating my company, Kaizen Assembly, to Bellingham, WA. I had grown up there but had spent about four years in the South, living in Georgia and North Carolina. Kaizen Assembly was entering its second year and was growing like a weed. Our customer base had grown, and I was excited about continuing its success in the Pacific Northwest. It was around this time that I also began teaching lean manufacturing at the local technical school in its professional development program. This training outlet was and is a very exciting and nice complement to my company's current consulting services.

We received an e-mail from the school explaining that a local manufacturer was interested in obtaining training. More specifically, this company had seen the curriculum for our highly popular Kaizen and Kaizen Event Implementation course. At the time there was no class being offered in open enrollment format, so I contacted the person about going to the facility to meet and set up the training at the site. The company was Samson Rope.

A meeting was scheduled. Our new office in downtown Bellingham was in the process of being set up; the movers were unloading our furniture and supplies as I was heading out the door to meet with Samson Rope. Its facility was about 20 minutes from the office, so it was encouraging that a potential new client was down the street. Business travelers will appreciate this situation.

I arrived at the facility and met with the plant manager, Ken. We sat down and had a brief discussion of the kaizen course, the number of people involved, cost, and scheduling. Once those formalities were complete, I was taken into the manufacturing facility.

"Live" is the best description of what I first experienced. Hundreds of twisting, winding, and braiding machines moving rapidly created an extremely dynamic environment. It did not take long for me to determine what my potential approach would be after the training, as the processes involved lots of machinery, equipment, and automation. I will explain this later during the lean assessment portion of the chapter.

Ken explained that the company had just finished a very aggressive re-layout of the plant to incorporate individual work cells to reduce transportation, motion, and travel distance. Those moves prior to my arrival were really their first pass at lean, so they already had created a good foundation for future changes. Samson Rope had a large maintenance department of highly skilled employees who supported the operation on all three shifts, including any special projects that were needed to improve the facility.

The tour lasted about an hour, and we concluded our meeting with a quick discussion of the training requirements for room space, booklets, and AV equipment. It was a great start to our relationship.

Kaizen and Kaizen Event Implementation Training

The training was set for February, and the company had committed ten people to sit in on the four-hour class. This course is designed to create the catalyst for improvements and help develop the leadership and other managers into the force needed to support lean manufacturing.

- Kaizen event steering committee
- Kaizen champion
- Communication (boards, newsletter, and suggestion system)
- Kaizen event supply box
- Planning for kaizen events and the timeline
- How to conduct kaizen events
- Action items and follow-up

They were quiet at first, as most new students are, and I am sure there was some level of skepticism—all common and to be expected. Ken and some other employees had already participated in lean manufacturing training off-site in another state, and they left with doubts about its use in their plant. They were given a simulation during the training of an assembly line using single-piece flow to illustrate how a pull system works. It would have been a great example if those in attendance had a manual assembly operation. Samson Rope does not, so I understood their early doubts. I knew from the initial tour that I would not use this type of simulation, and they needed examples that they could connect with. Those examples would come later.

Once we broke the initial theory portion and got into the "how-to" part, the questions and comments began to formulate. I described in detail who is required on the kaizen event steering committee and their roles and responsibilities. We had a discussion about the importance of the kaizen champion and how they could distribute the responsibilities of this person among multiple people in the beginning. The communication system was outlined as well as the supply box and the kaizen monthly meeting. These are all familiar to you from reading this book. The Samson Rope people were convinced that good communication was needed to support future improvements and they were happy with the detail. We finished up with some final questions, and I thanked them for the opportunity to speak to them and wished them good luck on their lean adventures.

The Return Visit

A week or so after the training, I received a phone call from Ken in which he asked me to come back to the plant and discuss working together. He asked me for some brief information on how the consulting process works and we scheduled the meeting.

The following week I arrived once again at the Ferndale facility to discuss a potential partnership. Ken and I went over the details and agreed on a contract. The Samson Rope and Kaizen Assembly bond was created in late February 2007. I was to conduct a full lean assessment of the company, facilitate a strategy session, conduct a series of training classes, and facilitate a few kaizen events. It was a good start.

Lean Assessment and Strategy Sessions

I arrived in early March to conduct the lean assessment, which evaluates where a company is in regard to lean manufacturing and kaizen, but it also looks at other factors like communication, suppliers, customer base, and general information about the organization. The information gathered in the assessment provides direction on where and how to start, what projects the company should work on, training needs, etc.

The assessment took a whole day, and once it was complete, I had identified a few areas as potential starting points that would consume the rest of the year. From the assessment I concluded that the focus for the first year should be on the implementation of 5S, developing the company's kaizen program, and training the employees on the fundamentals of waste reduction, value-added and non-value-added work, performance metrics, 5S, kaizen, and kaizen events.

During our strategy session, we discussed how we could implement 5S plant-wide and complete it by the end of 2007. Considering that we were already in March and just beginning, I was a little hesitant about setting a deadline that tight. The group was excited about the opportunity despite some doubts about how it could impact the organization. In theory it appeared beneficial and doable.

We discussed training schedules and how to put together the company's kaizen program. The implementation mechanism for 5S would be through scheduled and planned kaizen events spread out through the year. Kaizen events alone would not be enough; as in any lean journey, there are only so many kaizen events a company can undertake. It was also decided that some of the supervisors would have to implement 5S on their own, outside of events. We would have to make sure that they had the resources and supplies to get it done. However, the majority of the implementation would be through kaizen events.

The group also decided on a goal of a 10 percent increase in productivity for the year—a modest start and a goal that was possible through 5S.

Samson Rope's Kaizen Program

We wasted no time and started to work on the internal program that would support all lean initiatives:

- Kaizen steering committee
- Communication
- Kaizen event supply box
- Monthly meeting

Kaizen Steering Committee

As I mentioned in previous chapters, every company is different in its organizational chart. Each company has to place the right people on the committee based on their roles in the company to ensure that all attributes of planning and conducting events are satisfied. At Samson Rope, the committee consisted of these people:

> Ken: Plant Manager
>
> Ray: Coating and Warehouse Supervisor
>
> Kevin: Facilities Project Manager
>
> Dan: Operations Analyst
>
> Bill: Large Rope Supervisor
>
> Billy: Purchasing
>
> Teresa: Work Cell Supervisor

We felt that the people selected provided a good representation of the company and could help make decisions that would support the company's lean initiatives.

Communication

Kevin took the lead on getting the communication boards purchased. I believe they bought two boards and placed one near the break room/company information area and another elsewhere within the facility. Both were very visible to the employees. Dan helped put together the suggestion box and forms. Samson Rope already had a nice system of communicating company information and performance in a room near the back entrance, so the suggestion box was placed there.

Kaizen Event Supply Box

Kevin and his maintenance crew began the construction of the kaizen event supply box. They made it out of wood and installed wheels on it so that it could be moved around as needed during the kaizen events. It was designed to have a top lid that could be locked and that allowed for a writing surface. The box was constructed nicely and as I was shown it for the first time, I thought, "This thing is going to be used a lot." It wouldn't look new for long.

Monthly Meeting

The newly formed kaizen steering committee scheduled its first meeting to discuss the training schedule and the first kaizen event. I was impressed with their motivation to get going. We agreed to start with an introductory course on lean that would address the concepts. It was intended for the whole company, so they rented a large conference hall at a local hotel and packed the room. Once that course was complete, we scheduled the kaizen course and the 5S course. Once we completed that first round of training, we were ready for the first kaizen event.

Kaizen Event 1, May 7–11, 2007: Cell 5, Cell 8, and Splicing

The committee decided on a three-team kaizen event to set the tone for the rest of the year. I felt it was aggressive, but a multiteam event is doable if the resources are available. Samson committed between five and seven people for each team and we went for it.

There was a lot of fanfare and communication for this first event. Day one started out as planned, with all three teams sorting through the cells looking for anything and everything that was not needed. The splicing team had a much larger area to work in. Kevin was the leader for that team and he knew he had his hands full. The splicing area was complex since the work was a manually intensive process of "splitting rope" and customizing it for customers. The work varied from one week to another; one job required very little material and supplies, but the next needed a work area in which to place material and supplies. Kevin and his team began sorting.

Teresa had cell 5 and her team started off well. Teresa was eager for 5S but was having a hard time taking the idea from training to incorporating it into her area, a perfectly normal response to 5S.

Bill was leading the cell 8 team. Cell 8 was where the larger rope in the facility was made, so his goal was floor space reduction to provide better transportation in and out of the cell. 5S can be challenging when the product manufactured is big and naturally consumes floor space. Like the other teams, his team started the sorting portion nicely.

As day one slowly came to an end, the teams had done what I had asked of them: completed sorting by the end of the day. All teams had successfully removed a variety of things into a designated red-tag area for review. Day one ended with a quick team meeting to discuss progress and to see if we needed to shift people around or get more supplies. Kevin had supplied the event box well, and so far we were looking good. Kevin felt that his team might need to sort more on day two since it was a larger area. I could tell that the splicing team needed some extra help, but I wanted to see how Kevin would react to the amount of work still to do. There were a lot of items to set in order in splicing. Rather than interfere, I wanted to gauge Kevin's project management abilities.

Set in Order was the name of the game for the next two days. Since the work cells used mostly mechanical equipment, there was enough oil and dust on the floor that the team would have to use spray paint, rather than floor tape, for floor identifications. Floor tape was still used in some areas, though. Each team began identifying the items that would go on tool boards. Bill's team wanted to put new aisleways in the cell for improved flow, so some of his team members began to lay down masking tape to mark where the aisles would go. They also began to identify where carts, garbage cans, pallets, and any floor items would be located.

Teresa's team began with similar approaches to Set in Order, including the organization of the supervisor's workbench that is in each work cell. Everyone knew that all items had to have a home with clear designations and identification.

Kevin's battle continued in splicing. He had torn the place apart and was slowly putting the area back together. His approach was good, as I always encourage my 5S teams to clear the area and start from scratch. His team appeared to be overwhelmed, however, and I felt compelled to remove my consultant hat and put on the helper hat. I asked Kevin how I could help. He quickly gave me a broom and a vacuum to start the Scrub/Shine portion. Ken, the plant manager, started to come by the three areas to get a feel for where the teams were, and we discussed progress. At this point he was content but made comments about the mess in splicing. I explained that there were no major red flags at that point.

The teams continued throughout the day organizing, painting, and labeling. The same activities rolled into the third day. Cell 5 and cell 8 were moving along fine, to the point where I started to ask their team leaders to divert resources to the splicing team. On the third day I was getting nervous and Kevin was starting feel the same. Kevin and I looked at the team's progress and quickly reassigned people to specific tasks to try to get caught up. Ken had made another pass through the area and was now getting concerned that the team would not finish in time. I had complete confidence in Kevin's ability to manage his team; now it was a matter of "hands." By the end of day three, we were able to pull some more people off the other two teams to help the splicing team. Cells 8 and 5 were nearing completion. The first signs of 5S were beginning to appear as masking tape was removed after the paint had dried, stencil markings on the floor were complete, and tool boards were starting to be installed. It was happening.

Day four was now all about the splicing team. The other teams were still working hard to finish up, so we focused on Kevin's team. As in a lot of kaizen events, sometimes it feels as if there is too much to do with no end in sight. And then, almost magically, something happens: The team unites even more and *boom!* It gets done. This is what happened on Kevin's team. By the end of day four, the splicing area looked great and the other two teams were nearing completion.

We had an end-of-day meeting to discuss any final work to be completed on day five and to describe to the teams how to put together the report-out presentation. All team members were tired but happy, and a sense of accomplishment was prevalent.

The teams came in on day five and finished what they could, knowing that some amount of unfinished work would go on an action item list. The kaizen event ended with a great presentation by the three team leaders. The company had shown up in force, and they packed the break room for the first kaizen event report out. The teams provided a tour for the employees and a lot of questions were asked. Mostly, there were a lot of shocked faces and expressions. 5S had begun—with more to come.

5S Continued

It was back to work for the people at Samson Rope, as they entered the most difficult part of the process: sustaining and continually improving. The kaizen steering committee scheduled another meeting to discuss the development of the 5S audit and tracking system. They developed a preliminary 5S audit form and an audit schedule and posted a tracking sheet. The tracking sheet was placed in the company's information area and near each work cell. I explained that they should start auditing only those areas that had had the 5S implementation. We also discussed the next kaizen event and how 5S would be implemented in other areas outside of events. Teresa was the supervisor for three work cells, and she now had experience with 5S as a kaizen event team leader, so she committed to implementing 5S with her employees in her other areas of responsibility, on her own. Teresa was quickly becoming the 5S champion at Samson Rope. We had faith that she would forge ahead.

The committee encouraged the plant to look at what the teams did on the first event and begin to apply 5S. The culture of change was beginning.

The next official kaizen event was scheduled for September, and the area selected was the maintenance department.

Kaizen Event 2, September 24–28, 2007: Maintenance

Kevin was selected as team leader for this event since he was the supervisor of the facility project manager. The committee also selected Mike, who was the maintenance supervisor, to co-lead the event. Kevin wanted to break the team up into two sub-teams and have one focus on the actual maintenance work area and another on the part/supplies room. I thought it was a good approach. Maintenance 5S events bring different challenges since the maintenance department is a community area where people share tools and supplies depending on the need for the given day. These departments are often a dumping ground for the plant, and the maintenance staff also likes to hold on to things for what-if situations.

So we began. The sorting process was less crazy than I was anticipating. It was not the first time in my journey with Samson Rope that I found their resistance to change to be very low. They found the whole process of sorting out maintenance to be somewhat invigorating. Old tools, machinery, fixtures, parts, bins, and shelves were all removed. The red-tag area for this event was filling fast. There were three distinct areas in the maintenance department: the actual workbench/tools area, a parts room, and a small machine shop. Like a lot of other maintenance areas, it was dirty from all the work that they do, so I knew that cleaning would get us only so far. Paint, paint, paint would be the name of the game.

By the end of day one, the team had accumulated a very large pile of stuff. A significant amount of money was connected to those items, but there was no love lost there; if something was not used anymore, it left.

The real fun began on day two. Set in Order would be taken to a whole new level as this team was to seriously raise the bar on designations and identification. The tools boards were to be monstrous. The sheer number of tools needed to support the hundreds of machines in the plant was overwhelming, even after some healthy sorting. It was a three-shift operation with multiple employees; this department needed extra sets of tools in some cases. They had already started a tool board prior to the kaizen event after witnessing what went on during the first event in May, but there were still two to be made.

Each board was painted black to make the outline of the tools stand out. Each tool was marked based on the board to which it was assigned (i.e., board 1, board 2, etc.). Labels were placed above each tool, and everything had a home. Completing the tool boards took time, so Mike assigned a small group to focus just on them.

Other team members were organizing the cabinets that were used to hold larger, more expensive tools. I asked the team to seriously consider removing the doors to continue the theme of visibility. They decided to keep the doors because the items inside were expensive. I continued the fight but eventually gave in. It is important to learn to strike a balance between applying lean and giving a little to ensure that change is accepted. I compromised by asking that they put labels on the cabinet doors indicating what the contents were.

Dan, the operations analyst for Samson Rope, seemed to take on a common role with the kaizen events. He took part in the previous event and was also part of this project. He seemed to become the dirtiest of all team members on every event. Dan cleaned, sorted, and volunteered for a lot of the manually intensive work, often covering himself in dirt and grease. His contributions to the lean journey would become a huge asset. I helped Dan finalize locations for various tools and supplies on shelves and racks.

The team pushed hard into the third day, spray-painting equipment to give it a showroom look, painting the floor, stenciling, and organizing the parts room.

Kevin stayed focused on this parts room, organizing items based on use and labeling everything so people could find the items they needed faster. The maintenance department was beginning to look like something other than a maintenance department. Toolboxes and tool chests were disappearing. Drill presses and other machines in the work area that were previously out in the open, disrupting movement, now could be placed out of the way because of the floor space being opened up. Cleaning supplies, brooms, dustpans, paint cans, mops, jacks, etc., were all going vertical on boards or getting new homes. Ken made his usual walk-through and was happy with the progress.

As in all kaizen events, we concluded the event with another great report-out presentation and a tour. The lean journey was going in the right direction at that point.

Completing 5S

With the fourth quarter now within view, the sense of urgency picked up to get the 5S initiatives done. There were still a couple of areas to finish. Teresa had successfully completed 5S in her other cells with the help of her workers. Her enthusiasm for 5S was great, and we did not have to worry about her areas anymore. Dan and Kevin had also been working on the 5S audit system and began sending me the results of the weekly audits on all shifts. As in all journeys, there was some level of resistance to having to put things back all the time, but it was minor.

People began to notice the increased productivity, reduced walking, and reduced confusion we were seeing with this first pass at lean.

We scheduled a kaizen steering committee meeting to discuss the last new areas to implement 5S. Rather than having a kaizen event, which really did not appear to be needed, we agreed to take a group of people for one day and just get it done. I volunteered my time to help.

The groups spent a whole day finishing up the areas. As the groups finished, they decided to schedule one more event. What remained was the coating department that was supervised by Ray. This event was slated for December.

Kaizen Event 3, December 3-7, 2007: Coating

We had reached the last month of the year and had just one area to complete. Ray had initiated 5S in the warehouse as well, but the coating area required its own event. As a consultant, I have to make the tough decisions about what each company can begin next. We had focused purely on 5S for most of the year and did not incorporate any other waste reduction technique. It was at the coating event that I wanted to try some new approaches. Samson Rope had hired an intern from the local university, and we had conducted time studies and analysis to provide the team with ideas on cycle times, machine downtime, and operator walking. Basically, this information would be used to make more improvements to flow outside of 5S. I had to see if they were ready. The information collected showed a lot of opportunities to reduce waste. Based on this information, Ray had come up with a new layout idea that would help flow incoming and outgoing WIP.

Sorting was light in this event and the Set in Order phase began quickly. Half the team worked on 5S, and by now it was like old news to them. They had become solid veterans of 5S. Another group worked on a signal system to help the workers know what rope to coat in which order, based on color and deadlines. The 5S team marked off a holding zone for incoming WIP that would then be placed into a queue in a designated lane for a specific coating machine. This would allow the supervisor to place things in order as needed, and the workers could simply work on the items in the queue, reducing confusion.

A lot of great ideas and techniques implemented to improve flow. I wish I could go into more detail, but I am keeping the information here to a minimum since some of it is proprietary.

The event moved along fine as we watched the coating area become the last piece of the 5S puzzle. The area was completely changed to accommodate better flow, and the 5S implementation was complete.

2007 Complete

The 2007 lean journey with Samson Rope was quite amazing. The company saw record improvements in productivity, output, and delivery. 5S was not the only contributor to its success as other aspects of the business improved as well. As the year ended the Lafayette, LA, plant was beginning to get anxious about when it would happen for them. It was at this point that the alignment of the organization's 5S programs would begin.

Lafayette, LA: Lean Assessment

As Ferndale began frequent auditing and moving on to setup reduction and other waste reduction efforts, I hopped on an airplane in the last week of January 2008 and headed to meet the Lafayette employees. Ron, the plant manager in Lafayette, had already been to the Ferndale plant and seen the improvements made there. Ron, Mark (Samson Rope vice president of operations), and I had discussed starting the lean journey in Lafayette in 2008 with the intention of implementing 5S as the starting point. I still wanted to conduct my assessment to see if there were other important elements to address.

I met with Mark and Paul, the new purchasing manager (Ferndale). They were going down for other reasons, but Mark wanted to show his support for this new initiative. I conducted my day-long assessment to provide the company with a lean baseline; one of its findings was the need for 5S.

To help expedite the process and to get the plant 5S-compliant by November 2008 in preparation for new products coming in, we scheduled a full intensive workshop to get the frontline supervisors ready. To maintain some consistency, the training was similar to the sessions conducted in Ferndale.

I spoke considerably about waste, floor space reduction, 5S, and kaizen events. I also included total preventive maintenance training. Mark was eager to get moving with 5S. It was needed in the plant and he had a vision of helping Lafayette become a showroom facility. As at Ferndale, we wasted very little time and scheduled the first kaizen event. Ron, Mark, Paul, the now newly formed Lafayette kaizen steering committee, and I assessed potential areas. Since I had not spent a lot of time there, I suggested a three-day kaizen event to get our feet wet. We picked the Large Rope area called 901. It would also involve a small splicing area. The Lafayette kaizen steering committee consisted of the following people:

Ron: Plant Manager

Byron: Purchasing

Lisa: Safety/Quality Supervisor

Dave: Engineering

Joann: Shipping/Receiving Supervisor

Chris: Large Rope Supervisor

Terry: Maintenance Supervisor

Kaizen Event 1, February 20–22, 2008: Large Rope 901

Since the Ferndale plant now had ample experience with 5S and kaizen events, Dan and Janet (the controller) came down to help out on Lafayette's kaizen event. The kaizen steering committee selected a nice mix of employees to be on the team. Like the Ferndale events, this event involved line workers, machine operators, maintenance, quality, and the office. It was going to be a short event, so it had to move quickly. Sorting in a three-day event needed to be finished by lunch on the first day, so that we could quickly get to Set in Order. Chris, the Large Rope supervisor, was selected as the team leader for this event.

We got right to sorting and opened all cabinets, drawers, tool chests, and storage units. Dan, from Ferndale, continued his role as the dirt man; he climbed up on the large machines in the Large Rope area and began to clean and blow out the dirt that had accumulated over time. Everyone else started identifying unnecessary items and placing them in a red-tag area. The process was moving fast—so fast that the team had to go to the outside garbage cans and pull out things that did not get evaluated long enough in the red-tag area; there were some things that needed to go to maintenance. This was somewhat comical as far as I was concerned, and it does happen when people get excited.

We flew through the sorting and met up for lunch to discuss progress. Most of the sorting was complete, and we started the Set in Order phase right away. Byron focused on cleaning machinery and applying new paint. Other team members began tool board construction, trying to use the same approach as the Ferndale maintenance team had. People were marking off the floors and making homes for all supplies. The 901 area had a community gear supply in a cabinet, and a lot of walking back and forth was necessary during the day to set up machines. So the operators on the team identified those gears needed at each machine and installed gear racks right next to the machine controls to eliminate wasted motion. Everything was painted, and I mean *everything*.

Dan continued his cleaning work. Janet from Ferndale focused on labeling. Chris was a great team leader. His ability to manage the crew and keep people focused was exceptional. I had to intervene a few times but the incidents were minor in nature.

The team had cleared out the small splicing area and made some changes to layout to improve overall flow for people and material. Incoming and outgoing locations for WIP were located and tool boards were getting made. We pressed on into the second day. A level of competition was quickly forming between the plants as the Lafayette crew wanted to really show their stuff. They had an equal amount of enthusiasm for what was to be done.

It was also important to recognize the work they had done on their kaizen supply box. Marty, one of the maintenance employees, built probably the biggest supply box I have seen, and it was loaded up for the event. However, the committee underestimated the amount of paint needed, necessitating a couple of runs to the local hardware store. Again, this is a common occurrence during kaizen events.

We wrapped up the kaizen event on the third day with few to no action items. It was a great start for them, and they were excited about the next kaizen event. Before heading home, I spent some time with Ron discussing the pros and cons of the event and the expectations of his newly formed committee. Ron was giving full support and we scheduled a conference call to be conducted during their next kaizen meeting. I boarded a plane the next morning and headed home.

Kaizen Event 2, April 7–11, 2008: Area G, Area D, Area B

I called the Lafayette plant about two weeks later to listen in on the kaizen steering committee meeting. We discussed the previous event, progress, lessons learned, and the next kaizen event. Because of the success of the event and because they had maintained the 901 area from the previous event, we scheduled a multiteam event that would take a whole week. The committee selected three areas and picked the team leaders and members. The event was scheduled for early April, and Kevin and Teresa from Ferndale volunteered to help as well. Ken also was coming on other business.

We showed up in force and we started day one with the usual activities. In preparation for the event, a lot of sorting had been conducted, as the people in Lafayette were getting excited about improvement and could not wait. That type of enthusiasm is great and I always encourage people to sort early if they want. On day one, Lisa, the safety and quality supervisor, came up to me and asked me to come see her department. She had already started on her area and was in the process of implementing 5S in her spare time with the help of her employees. I was happy see other activities going on outside of kaizen events. Motivation was high in Lafayette.

The teams worked hard sorting and moved into Set in Order by the end of day one. Teresa and Kevin from Ferndale became co-facilitators, helping the team leaders find their way. They did not need to help too much, but when they did, they made decisions as needed.

As in the first event, masking tape was used to outline aisles and floor locations. Equipment was getting cleaned and painted. Gears were being pulled from community holding areas and being placed right at the machines for point-of-use applications. Tool boards for tools and cleaning supplies were getting constructed and slowly being filled with the necessary items. Ron, the plant manager, was seen a lot on the floor, assessing the progress and showing his support for the project.

There were no real stoppages, just a few redirects here and there to keep things moving. As kaizen events near the end, people need to be working on important items. For instance, if the team is in day four and people are wiping down hand trucks, but no tool boards are up, the event leader must redirect. It happened a few times on this event but there was nothing major. It was the second event at Lafayette and the team leaders were learning as they went, and doing very well. As the event was nearing its end, some of the teams were finishing early, and we shifted people around to help other teams that had not yet completed their work.

We concluded with another energetic report-out presentation, and I expressed my satisfaction with their progress. It was a fun group to work with and I was excited about my next return.

Samson Rope Progress: Ferndale and Lafayette

I added this part of the chapter near my submission to Prentice Hall. Kaizen events continued in the Lafayette facility. I went down for two more events, one in June and one in August of 2008. Each one was as successful as the previous events had been. Lafayette was able to complete the 5S implementation plant-wide by the end of the year. Additional training was also conducted to present possible next steps. We trained on total preventive maintenance, setup reduction, time studies, process mapping, and quick changeover.

The Ferndale plant spent 2008 learning about value stream mapping, time studies, and increasing uptime, and we conducted a few kaizen events to further reduce waste, create visual production boards, and look at material replenishment systems. Ferndale also refined its 5S auditing system, which incorporated a rotational auditor program, and quarterly incentives for the highest 5S scores. Outside of this great story, both plants had challenges, setbacks, and struggles as in any lean journey. Lean is truly a battlefield and it can be harder once 5S is complete.

Both plants have to balance running the business and conducting continuous improvement. Lean is about two steps forward, one step back. Resistance will appear at every level along the way. Change is always tough, even for those organizations that have been practicing lean for many years. The bottom line is that Samson Rope has proved to itself that it can practice lean. And it is this continued push toward being the best that will keep them atop the rope manufacturing industry.

Other Samson Rope Employees to Recognize

- Eston
- Ron
- Mike
- Marlene
- Bill
- Asad
- Chris

I hope I mentioned all of you.

Conclusion

My goal in writing this book was to provide you with the foundational elements needed to support continuous improvement. The company kaizen program has proven to be very successful in giving organizations the vision, focus, and drive to make lean a part of doing business. It is also important to understand that taking lean in its "textbook" version is dangerous, as any journey must be custom-fit to the company's needs, processes, products, customer base, and employees. I hope that you will take the concepts of the kaizen steering committee, the kaizen champion, lean communication system, tracking, team selection, team leader selection, kaizen event facilitation, and many of the other ideas described here and make them fit the unique culture within your company. Only then can you succeed with kaizen and lean. I wish you the best of luck with your lean endeavors.

Chris Ortiz
Kaizen Assembly

Index

A

Action items
  5S, 84
  champion follow-up, 52–53
  Samson Rope Technologies, Inc., 119
  standard work events, 102–103, 106–107
  tracking, 31–33
Alternatives to champions, 53–54
Analyzing collected data, 64
Area selection, 70
Assembly lines
  standard work events design, 102–104
  time and motion studies, 93
Audit system
  5S events, 88–89
  Samson Rope Technologies, Inc., 120

B

Balancers, tool, 105
Balancing, line
  reviewing, 100–101
  rules, 97–98

Benefits of kaizen, 7–8
Budget
  managing, 47
  tracking, 30–31

C

Cabinet limitations, 80
Capturing work content, 95
Case study. *See* **Samson Rope Technologies, Inc.**
Champions, 39–40
  alternatives, 53–54
  choosing, 47–49
  costs, 49–50
  need for, 40–41
  overview, 24–25
  purpose, 54
  responsibilities, 50–53
  skill sets, 41–47
Changeover skills for champions, 45
Classrooms for training, 50
Cleanup procedures, 88
Closing out standard work events, 110
Coating department in Samson Rope Technologies, Inc., 123
Collected data, analyzing, 64

Collecting data
  5S events, 77
  current state, 44, 64, 73
  time study, 94
Color for tower lights, 109
Communication
  common mistakes, 9–10
  developing, 33
  newsletters, 34–35
  Samson Rope Technologies, Inc., 116–117
  suggestion boxes, 35–37
  systems updating, 61, 72
Communication boards
  champion responsibilities, 51
  description, 34
Company cultural change, 4
Company kaizen programs, 17–18
  champions, 24–25
  communication, 33–37
  event steering committees. *See* Steering committees
  tracking. *See* Tracking
Completing Samson Rope Technologies, Inc., 122–124
Conference rooms, 63
Costs of champions, 49–50
Cross-functional teams, 47
Current productivity evaluation, 57
Current state, collecting, 44, 64, 73
Customer complaint information, 57

D

Daily walk-throughs, 88
Data collection skills for champions, 43–45
Date tracking, 28
Dead space, 79
Deadlines
  action items, 32, 84
  red-tag removal, 78
Design ideas, 64
Distance, travel, 14–15
Diversity in teams, 47
Documenting work content, 94–96

E

Effective time in standard work events, 92
End-of-day cleanup procedures, 88
End-of-day meetings, 99, 102, 119
Engineers
  responsibilities, 5, 17
  on steering committees, 19–20

Errors, eliminating, 13
Estimating spending, 59–60
Event-lean mode, 8
Event supply box
  5S events, 80
  constructing, 71
  Samson Rope Technologies, Inc., 117
  for supplies, 60, 63
Events overview, 8–9
  5S. *See* 5S events
  champion skills, 43
  common mistakes, 9–11
  metrics, 12–13
  schedules. *See* Schedules
  selection, 26–27
  standard work. *See* Standard work events
  steering committees. *See* Steering committees
  tracking. *See* Tracking
External options for champions, 48–49

F

Facilities managers on steering committees, 22
Finalizing team members, 62–63, 72
5S events, 69
  area selection, 70
  audit sheets, 88–89
  champions for, 42–43, 54
  communication system updates, 72
  current state information gathering, 73
  incentive programs, 90
  maintenance, 84–85
  meeting space selection, 72
  outside assistance, 72
  plant and general manager meetings, 74
  Set in Order and Scrub events, 79–83, 86–87
  Sort events, 74–79, 85–86
  spending, 71
  Standardize events, 83–84
  supplies, 71, 74
  sustaining tips, 87–90
  teams. *See* Teams
  tracking sheets, 89
  updating supplies and outside resources, 72–73

Floor space reductions, 13–14
Floor taping and designations
  5S events, 80, 83
  standard work events, 108
Follow-up for action items, 52–53
Food preparations, 66

G

Garbage cans, 79
Gathering data
  5S events, 77
  current state, 44, 64, 73
  time study, 94
General managers
  meeting with, 65, 74
  on steering committees, 19
Goals, 11
  establishing, 59, 71
  tracking, 30
Green color for tower lights, 109

H

Highest sample removal from time studies, 96
Hiring costs for champions, 49–50
Human resource managers on steering committees, 21–22

I

Incentive programs, 90
Internal options for champions, 48
Internal quality information, 57
Inventory, 12, 98
Investments in champions, 49

K

Kaizen Assembly: Designing, Constructing, and Managing a Lean Assembly Line, 34, 111
Kaizen overview
  benefits, 7–8
  leaders, 5–7
  overview, 4
  people, 5
Kickoff meetings, 98–99

L

Leaders, 5–7
  champion responsibilities, 52
  selection, 28, 58–59, 70–71
  standard work events, 103
  tracking, 28

Lean overview, 3–4
  champion skills, 42
  leadership, 5–7
  Samson Rope Technologies, Inc., 115, 124–125
  transformation to, 5–6
Length, tracking, 28
Lessons from a Lean Consultant, 6
Line balancing
  reviewing, 100–101
  rules, 97–98
Line designs
  champion skills for, 46
  standard work events, 102–103
Line operator input, 35
Lowest sample removal from time studies, 96

M

Machine shop projects, 104
Maintenance events
  5S, 84–85
  Samson Rope Technologies, Inc., 120–122
  standard work events, 102, 104
Maintenance personnel on steering committees, 22
Managers
  improvement responsibilities, 5
  meetings with, 65, 74
  on steering committees, 19, 21–22
Manufacturing engineers
  responsibilities, 5, 17
  on steering committees, 19–20
Materials
  champion skills for, 46
  organizing, 83
Materials managers on steering committees, 22
Measurable improvements, 7–8
Meetings
  end-of-day, 99, 102, 119
  kickoff, 98–99
  with managers, 65, 74
  midday, 103–104
  monthly, 31–33, 51, 117
  Samson Rope Technologies, Inc., 112–113
  space selection, 72
  team, 64, 73–74

M

Metrics
  floor space, 13–14
  inventory/WIP, 12
  productivity, 12
  quality improvements, 12–13
  travel distance, 14–15
  workstations, 14
Midday meetings, 103–104
Monitoring champion responsibilities, 53
Monthly meetings
  champion responsibilities, 51
  Samson Rope Technologies, Inc., 117
  for tracking, 31–33

N

Newsletters
  champion responsibilities, 51
  description, 34–35
  generating, 61, 72
Non-value-added work, 62
Nonmeasurable improvements, 7–8

O

Open action items
  champion follow-up, 52–53
  standard work events, 106
  tracking, 32–33
Operations managers on steering committees, 21
Operators
  data collection phase, 95
  input from, 35
  on steering committees, 24
Ordering supplies, 60–61
Outside assistance, scheduling, 61, 72
Outside resources, updating, 63, 72–73

P

Painting, 80–81
Parts organization, 83
Parts rack signs, 107–108
People of kaizen, 5
Performance evaluation, 57
Planning events
  common mistakes, 10
  scheduling, 33
Plant managers
  meeting with, 65, 74
  on steering committees, 19

Pre-event goals, 30
Preplanning
  standard work events, 92–98
  tracking, 29–30
Presentations
  5S Events, 84
  standard work events, 110
  tools, 104–105
Previous results, meetings for, 32
Process analysis, 93–98
Process/department/work area selection, 56–57
Process-oriented cultures, 4
Production managers on steering committees, 21
Production supervisors on steering committees, 23
Productivity
  evaluating, 57
  improvements, 12
Project management skills for champions, 47
Purchasing managers on steering committees, 22

Q

Quality improvements, 12–13
Quality managers on steering committees, 20
Quick changeover skills for champions, 45

R

Rapid improvement projects, 8
Red color for tower lights, 109
Red-tag campaigns, 75–77
Red-tag removal procedure
  deadlines, 78
  standard work events, 101–102
Red-tag sorting activity, 77–78
Red-tag teams, 99–100
Removal, red-tag, 78, 101–102
Report-out presentations
  5S events, 84
  standard work events, 110
Responsibilities
  champions, 50–53
  tracking, 31
Results
  monthly meetings for, 32
  tracking, 30
Retractors, tool, 105
Room selection, 63

S

Sales and output evaluation, 56–57
Samples in data collection, 96
Samson Rope Technologies, Inc., 111–112
  area events, 127–128
  audit and tracking system, 120
  cells and splicing, 117–120
  coating, 123
  communication, 116–117
  completing, 122–124
  event supply box, 117
  implementation training, 113–114
  initial meeting and tour, 112–113
  Large Rope area, 125–127
  lean assessment, 115, 124–125
  maintenance, 120–122
  monthly meetings, 117
  progress, 128–129
  return visit, 114
  steering committee, 116
  strategy sessions, 115
Schedules, 55–56
  collected data analysis and design ideas, 64
  communication system updates, 61
  current state information gathering, 64
  events, 33
  food preparations, 66
  outside assistance, 61, 72
  plant and general manager meetings, 65
  process/department/work area selection, 56–57
  room selection, 63
  spending estimates, 59–60
  supplies, 60–61
  supply placement, 64
  teams. *See* Teams
  updating supplies and outside resources, 63
  waste analysis, 61–62
Scrub event
  S5 events, 79–83, 86–87
  standard work events, 102, 104
Selection
  areas, 70
  champions, 47–49
  events, 26–27
  leaders, 28, 58–59, 70–71
  meeting space, 72
  process/department/work area, 56–57
  rooms, 63
Set in Order events
  S5, 79–83, 86–87
  Samson Rope Technologies, Inc., 123
  standard work events, 102
Setup reduction skills for champions, 45
Seven deadly wastes
  5S events, 71
  champion skills for, 41–42
  evaluating, 57
  non-value-added, 62
  Samson Rope Technologies, Inc., 117
  standard work events, 98
Shadow boards
  5S events, 81–83, 86
  standard work events, 105, 109
Shift walk-throughs, 88
Shine function, 104
"Shock and awe" effect, 8
Signs, 107–108
Skill sets for champions, 41–47
Sort events
  5S events, 74–77, 85–86
  red-tag areas, 77–78
  Samson Rope Technologies, Inc., 123, 125–127
Spaghetti diagrams
  for area selection, 70
  for current state, 45
Spending
  5S events, 71
  estimating, 59–60
  tracking, 30–31
Splicing in Samson Rope Technologies, Inc., 117–120
Standard work events, 91
  action items, 102–103, 106–107
  closing out, 110
  end-of-day meetings, 102
  floor taping and designations, 108
  kickoff meetings, 98–99
  line designs, 102–103
  maintenance and Machine shop projects, 104
  midday meetings, 103–104
  preplanning, 92–98
  red-tag teams, 99–100

Standard work events (cont.)
　Scrub/Shine function, 104
　shadow boards, 105, 109
　subassemblies, 105–106
　subassembly build levels, 108
　tool presentation, 104–105
　tower lights, 109
　visual management, 106–107
　workstation and parts rack signs, 107–108
Standardize event, 83–84
Stations
　overview, 14
　signs, 107–108
Status tracking, 31
Steering committees, 18–19
　human resource management, 21–22
　maintenance and facilities management, 22
　manufacturing engineering management, 19–20
　operations and production management, 21
　operator representatives, 24
　plant and general management, 19
　production supervisors, 23
　purchasing and materials management, 22
　quality management, 20
　Samson Rope Technologies, Inc., 116
Stopwatches for time and motion studies, 94
Straighten process, 78
Strategy sessions for Samson Rope Technologies, Inc., 115
Subassemblies, 105–106, 108
Suggestion boxes
　champion responsibilities, 52
　description, 35–37
Suggestions evaluation, 57
Supplies
　5S events, 71
　ordering, 60–61
　organizing, 83
　placing at gathering space, 64, 74
　updating, 63, 72–73
Surfaces, evaluating, 79
Sustaining tips for 5S events, 87–90

T
Takt time, 92–93
Taping work areas
　5S events, 80
　standard work events, 108
Team leaders
　champion responsibilities, 52
　selection, 58–59, 70–71
　standard work events, 103
　tracking, 28
Teams
　common mistakes, 10–11
　cross-functional and diverse, 47
　finalizing, 62–63, 72
　goals establishment, 59, 71
　meeting with, 64, 73–74
　sorting, 75–77
　spending time in chosen area, 63, 73
　tentative lists, 57–58, 71
　tracking, 28–29
Tentative team member lists, 57–58, 71
Time, line balancing by, 97
Time and motion studies, 62
　champion skills for, 44
　standard work events, 93–97
Tool balancers, 105
Tool boards, 81–83, 86
Tool chest limitations, 80
Tool presentation, 104–105
Tool retractors, 105
Touch time, 92
Tower lights, 109
Tracking, 25–26
　5S events, 89
　action items, responsibility, and status, 31
　budget and spending, 30–31
　champion responsibilities, 52
　date and length, 28
　event selection, 26–27
　event team leaders, 28
　monthly meetings for, 31–33
　pre-event goals, 30
　preplanning and preplanning responsibility, 29–30
　results, 30
　Samson Rope Technologies, Inc., 120
　team members, 28–29

Training responsibilities, 50–51
Training rooms, 63
Travel distance, 14–15

U

U-shaped cells, 102–103
Updating
    communication systems, 61, 72
    supplies and outside resources, 63, 72–73

V

Value-added work, 62
Value stream mapping (VSM), 44, 62
Visibility
    5S events, 86
    champion skills for, 42–43
    standard work events, 106–107
    tools, 79–81
Volume requirements, 92

W

Walk-throughs, 88
Waste analysis, 61–62
Waste elimination, 4. *See also* Seven deadly wastes
Work area selection, 56–57
Work content, documenting, 94–95
Work flow skills for champions, 46
Work in process (WIP), 12, 99
Work surface evaluation, 79
Worksheets
    champion responsibilities, 52
    tracking, 25–26
Workstations
    overview, 14
    signs, 107–108

Y

Yellow color tower lights, 109

LESSONS FROM A LEAN CONSULTANT

Published: March 2008
Author: Chris A. Ortiz
ISBN-10: 0131584634
ISBN-13: 9780131584631

LESSONS FROM A LEAN CONSULTANT

A nuts-and-bolts approach to making lean manufacturing work through transforming the company's culture

The concepts of lean manufacturing are the answers to many problems factories face in this day of increased global competition. The improvements to productivity, quality, lead times, and cost are well documented. However, lean manufacturing is only successful if commitment and dedication are presented by those who run the business. Let expert Chris Ortiz show you what he has learned teaching Fortune 500 companies to become more efficient businesses, including:

- How to create a methodical approach to lean implementation
- What pitfalls might lie ahead, and how to avoid them by learning from mistakes made by others
- How to succeed using the kaizen philosophy of continuous improvement

CHAPTER 1:
Case Study: How Lean Failed

CHAPTER 2:
The Change Commitment

CHAPTER 3:
The Lean Infrastructure: Kaizen

CHAPTER 4:
Early Stumbling Blocks

CHAPTER 5:
Operator and Supervisor Involvement

CHAPTER 6:
Lean Training Programs

CHAPTER 7:
Lean Manufacturing as a Growth Creator

CHAPTER 8:
Lean Leadership Made Simp

APPENDIX A:
Quick Reference

APPENDIX B:
Supplemental Material

For more about this title, visit www.informit.com/title/0131584634

Register the Addison-Wesley, Exam Cram, Prentice Hall, Que, and Sams products you own to unlock great benefits.

To begin the registration process, simply go to **informit.com/register** to sign in or create an account. You will then be prompted to enter the 10- or 13-digit ISBN that appears on the back cover of your product.

Registering your products can unlock the following benefits:
- Access to supplemental content, including bonus chapters, source code, or project files.
- A coupon to be used on your next purchase.

Registration benefits vary by product. Benefits will be listed on your Account page under Registered Products.

About InformIT — THE TRUSTED TECHNOLOGY LEARNING SOURCE

INFORMIT IS HOME TO THE LEADING TECHNOLOGY PUBLISHING IMPRINTS Addison-Wesley Professional, Cisco Press, Exam Cram, IBM Press, Prentice Hall Professional, Que, and Sams. Here you will gain access to quality and trusted content and resources from the authors, creators, innovators, and leaders of technology. Whether you're looking for a book on a new technology, a helpful article, timely newsletters, or access to the Safari Books Online digital library, InformIT has a solution for you.

THE TRUSTED TECHNOLOGY LEARNING SOURCE

Addison-Wesley | Cisco Press | Exam Cram
IBM Press | Que | Prentice Hall | Sams
SAFARI BOOKS ONLINE

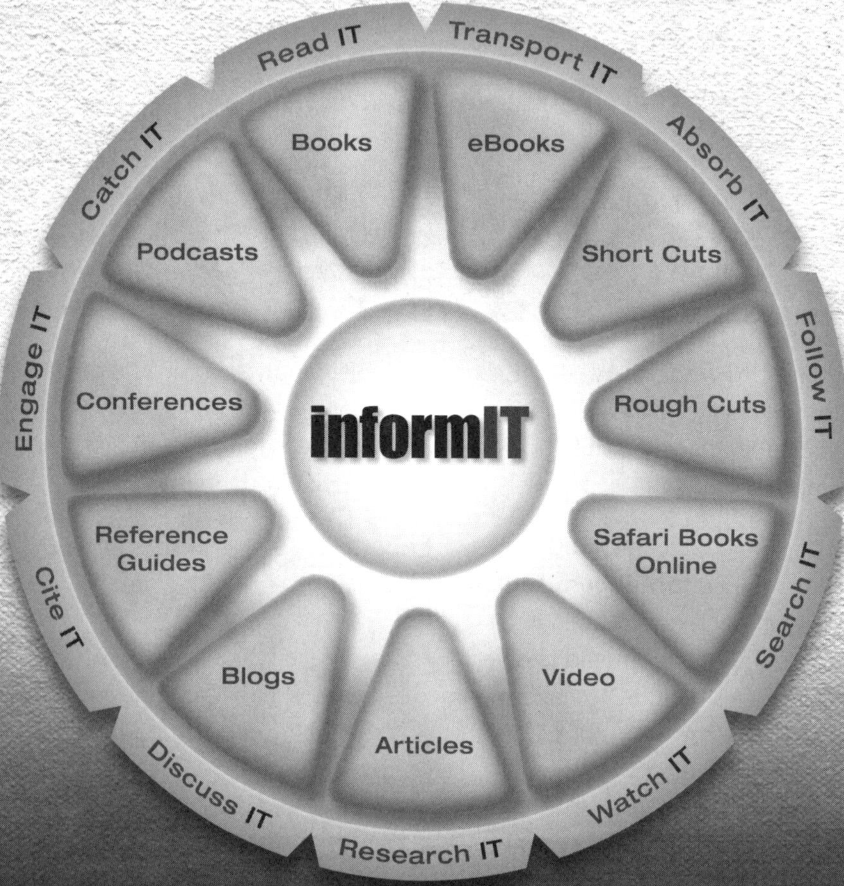

Try Safari Books Online FREE

Get online access to 5,000+ Books and Videos

FREE TRIAL—GET STARTED TODAY!
www.informit.com/safaritrial

Find trusted answers, fast
Only Safari lets you search across thousands of best-selling books from the top technology publishers, including Addison-Wesley Professional, Cisco Press, O'Reilly, Prentice Hall, Que, and Sams.

Master the latest tools and techniques
In addition to gaining access to an incredible inventory of technical books, Safari's extensive collection of video tutorials lets you learn from the leading video training experts.

WAIT, THERE'S MORE!

Keep your competitive edge
With Rough Cuts, get access to the developing manuscript and be among the first to learn the newest technologies.

Stay current with emerging technologies
Short Cuts and Quick Reference Sheets are short, concise, focused content created to get you up-to-speed quickly on new and cutting-edge technologies.

Your purchase of **Kaizen and Kaizen Event Implementation** includes access to a free online edition for 45 days through the Safari Books Online subscription service. Nearly every Prentice Hall book is available online through Safari Books Online, along with more than 5,000 other technical books and videos from publishers such as Addison-Wesley Professional, Cisco Press, Exam Cram, IBM Press, O'Reilly, Que, and Sams.

SAFARI BOOKS ONLINE allows you to search for a specific answer, cut and paste code, download chapters, and stay current with emerging technologies.

Activate your FREE Online Edition at www.informit.com/safarifree

> **STEP 1:** Enter the coupon code: UFSJMXA.

> **STEP 2:** New Safari users, complete the brief registration form. Safari subscribers, just log in.

If you have difficulty registering on Safari or accessing the online edition, please e-mail customer-service@safaribooksonline.com